Accidental
Medical
Discoveries

Accidental Medical Discoveries

How Tenacity and Pure Dumb Luck Changed the World

ROBERT W. WINTERS, MD

Skyhorse Publishing

Skyhorse Publishing books may be purchased in bulk at special discounts for sales promotion, corporate gifts, fund-raising, or educational purposes. Special editions can also be created to specifications. For details, contact the Special Sales Department, Skyhorse Publishing, 307 West 36th Street, 11th Floor, New York, NY 10018 or info@skyhorsepublishing.com.

Skyhorse® and Skyhorse Publishing® are registered trademarks of Skyhorse Publishing, Inc.®, a Delaware corporation.

Visit our website at www.skyhorsepublishing.com.

10 9 8 7 6 5 4 3 2 1

Library of Congress Cataloging-in-Publication Data is available on file.

Cover design by Rain Saukas
Cover photo credit: iStock

Print ISBN: 978-1-5107-1246-1
Ebook ISBN: 978-1-5107-1247-8

Printed in the United States of America

To Agnete, Charlotte, and George

Disclaimer

Nothing in this book should be construed as medical advice for the treatment of any condition. In addition, nothing in this book should be taken to represent the proper use of the medications and techniques described herein.

CONTENTS

Acknowledgments

This book has been nine years in the making. During that time, my wife, Agnete, my daughter, Charlotte, and my son, George, have been constant supporters of my work. Much of the writing was done in Denmark, my permanent home for the past eight years. The first and most deserving of thanks goes to my wife, Agnete, who read every word, and in doing so found errors in grammar and punctuation, and failure to be absolutely clear. George, my son, has from the very beginning provided me with ongoing critical support. I would have been hard-pressed to take on this project without his help.

I want to renew my thanks to Dr. Allan Pelch, my Danish doctor friend, who provided suggestions for the previous edition. Some months ago, I received a letter from a man named John Andersen, who is an attorney in Aalborg. His letter came as a complete surprise, since I had never met or spoken with the man. But I was happily stunned by his praise, which I ranked as a "rave review" of the self-published edition of the book. I responded with a letter of thanks, and through subsequent exchanges, Mr. Andersen has provided me with helpful answers to several problems with organization or language of the text. I feel fortunate in having John Andersen as an intelligent lay critic. I am also grateful in having a computer expert, Mr. Tonny Gumoes, who patiently and expertly tutored me in the mysteries of the computer.

The long distance between Denmark and an American medical library has proven to be much less formidable than it might seem, due to my long-standing relationship with the Shinn-Lathrope Health Sciences

Library of the Morristown Memorial Hospital in Morristown, New Jersey. Ms. Janina Kaldan and her staff have been reliable and quick in their retrieval of journal articles that I have needed. In my opinion, this library ranks at the top of the list of the many medical libraries that I have worked with during my long career.

Preface

My interest in the history of medicine was kindled many years ago when, as a first-year student at the Yale School of Medicine, I had the great good fortune to know and admire Professor John Fulton, a world leader in neurophysiology and a great scholar of medical history. I was especially stimulated by one of Dr. Fulton's anecdotes about the first use of penicillin in the United States.

Professor Fulton had a long, close friendship with Howard Florey that began with their student days at Oxford. Following Alexander Fleming's discovery, Florey led dedicated scientists at Oxford in the daunting task of isolating and purifying penicillin, while working under the immense handicaps imposed by early years of World War II.

In 1942, a Mrs. Ogden Ann Miller, the wife of the director of athletics at Yale, lay desperately ill with childbed fever in the New Haven Hospital. She was near death with a fever of 40° C. Her doctor, knowing of Fulton's friendship with Howard Florey, leader of the Oxford group that brought penicillin to the marketplace, pleaded with Fulton to arrange to get some of the then extremely limited amount of penicillin that was in the United States for Mrs. Miller. After a flurry of telephone calls, Fulton was successful in arranging for a small amount to be swiftly brought to New Haven. The wonder drug worked its miracle, and Mrs. Miller recovered completely. Indeed, she lived to the age of ninety, and her obituary appeared in *The New York Times*.

I had to wait a long time before being able to develop my growing interest in the history of medicine that was inspired so long ago by

Dr. Fulton. My professional life as a physician, extending over four decades, was fully occupied with the heavy demands of medical research and teaching, patient care, and the sometimes-onerous academic duties of a senior academic physician. Those years were followed by an additional ten-year stint as a busy entrepreneur in building a successful health-care company.

In retirement, I could at last devote long uninterrupted periods to the necessary research for writing this book. I first returned to the story of penicillin and to its chance discovery by Alexander Fleming in 1928. I soon learned that there was a great deal more to the simple, oft-repeated tale that Fleming obtained the mold from the bread of a sandwich. In fact, the role of the London weather was critical in creating the temperature conditions in the laboratory that were crucial for Fleming to make his discovery. Historically, Fleming is often singled out as the discoverer of penicillin, while Howard Florey and his colleagues are too often the unsung heroes of this tale.

Unraveling this tale has been a fascinating journey into the past, and it increased my appetite for other stories behind the discoveries of other drugs and devices. During the past eight years, I have researched the history of twenty-six medically important discoveries whose stories are presented in this volume. In telling these stories, I have limited my attention to the details and circumstances of the discoveries and have not attempted to deal with the present or future uses of the discoveries in modern medical practice.

The underlying theme of each of these tales is that the discovery was made by chance, often referred to as serendipity, which has played an important role in the discovery. Serendipity takes its name from Horace Walpole's description of a fairy tale entitled *The Three Princes of Serendip*. It seems that during the princes' travels, Walpole relates, that "the three highnesses were always making discoveries by accident and sagacity of things which they were not in quest of." He coined the word *serendipity* as meaning discoveries made by "accident and sagacity."

Since Walpole's time, there have been many different attempts to clarify or to define the word *serendipity*; nearly all emphasize "accidental" but many neglect "sagacity," i.e., wisdom. The 1936 *Oxford English Dictionary* (OED) defined serendipity as "the occurrence and development of events by chance in a happy or beneficial way." This is the way it is often used today. But the 1986 edition of the OED expanded the original definition of serendipity to include "a fact or an instance," but also a "faculty, ability, gift, habit, aptitude or talent."

In other words, serendipity is a property of the discoverer as well as of the event being discovered, which seems to confuse rather than enlighten. Robert K. Merton and Elinor Barber in their book *The Travels and*

Adventures of Serendipity defined serendipity as the fairly common experience of observing an unanticipated, anomalous, and strategic datum that becomes the occasion for developing a new or extending an existing theory. Louis Pasteur said it best when he wrote: *"Dans les champs de l'observation le hasard ne favorise que les esprits préparés."* ("In the fields of observation, chance favors only the prepared mind.")

Today, the use of the word *serendipity* has grown to the point of overuse and widespread abuse. For example, it appeared in newspapers and magazines some 13,000 times during the decade of the 1990s and in 636,000 documents on the World Wide Web in 2001. The word, once a "vogue word," has morphed into a "vague word" that is now in popular use as a "catchy" name for restaurants, cafes, boutiques, B&B's, inns, motels, skin care products, children's toys, a line of women's dresses, certain food products, yachts, and doubtless many other things. In this book, the author has refrained from adding any further usage and is quite content to use the term *accidental discovery*.

The book is neither a medical nor a pharmacology treatise. Rather, it provides insight, even perhaps amusement, into how some famous discoveries came about. They all occurred by chance, and they all occurred because some prepared mind grasped the significance of what had happened. My guiding principle has been to tell stories that are interesting, not only in the details of how the discovery was made, but also a little about the personalities of the discoverers.

Part I
Surgery and Anesthesia

The Gleam of Cold Steel: Surgery without Anesthesia

I began a scream that lasted unremittingly during the whole time of the incision.

—Fanny Burney, 1801

Introduction

In his autobiography, Charles Darwin wrote his reactions to observing two operations carried out without anesthesia: "I attended on two occasions in the operating theatre of a hospital at Edinburgh, where I watched two very bad operations, one on a child, but I rushed away before they were completed. Nor did I ever attend again, for hardly any inducement would have been strong enough to make me do so, this being long before the blessed days of chloroform. The two cases fairly haunted me for many a long year."

Surgery prior to the discovery of anesthesia was excruciatingly painful and often fatal. The operations that were performed were mainly amputations, removal of bladder stones, or draining of ovarian cysts. Speed was the key to the success of these procedures in order to minimize the agony of the patients under the knife. The clock was the measure of the surgeon's skill. An amputation by a skillful surgeon took less than two minutes. Achieving great speed became the hallmark of fame for both the public and for other surgical colleagues, whose speed was less than that of the fastest operator. Robert Liston, a famous surgeon, in setting out to best his own speed record for amputation of the leg, accidentally amputated one of his patient's testicles and two of his assistant's fingers.

During an operation, the frenzied patient provided a restless, screaming target for the surgeon crying out from excruciating pain. The surgeon was not immune to the suffering that he was inflicting nor was he insensitive to the dangers that a sharp scalpel might inflict on his writhing patient. This was an ever-present danger that Dr. Valentine Mott, a famous New York surgeon, wrote: "How often, when operating in some deep, dark wound, have I dreaded that some unfortunate struggle of the patient would deviate the knife from its proper course, that I would involuntarily become the executioner, seeing my patient perish in my hand?"

Fanny Burney's Surgery

During the early 1800s, Fanny Burney, a socially prominent English novelist, had been troubled for several years with recurrent pain and inflammation in her right breast. Her physician had diagnosed cancer and recommended surgical removal. Fanny consented even though no anesthetic was available: "Everything convinced me . . . that only this experiment could save me."

Fanny and her husband, Alexandre d'Arblay, a retired general, were living in France. Baron Dominque-Jean Larrey, an eminent military surgeon who served with distinction under Napoleon, agreed to perform the operation. Fanny recorded her agonies during the harrowing operation in what has been called "the most extraordinary piece of 'reminiscences' ever committed to paper." Some excerpts from Fanny's account follows:

M. Dubois placed me on a mattress and spread a cambric handkerchief upon my face. It was transparent, however, and I saw through that the seven men and my nurse instantly surrounded the bedstead. I refused to be held; but when, bright through the cambric, I saw the glitter of polished steel—I closed my eyes.

Yet, when the dreadful steel was plunged into the breast—cutting thorough veins, arteries, flesh, nerves, I needed no injunction to restrain my cries. I began a scream that lasted unremittingly during the whole time of the incision—and I almost marvel that it does not ring in my ear still, so excruciating was the agony! When the wound was made, and the instrument withdrawn, the pain seemed undiminished for the air that suddenly rushed into those delicate parts felt like a mass of minute but sharp and forked poniards [daggers] that were tearing at the edges of the wound. Yet again all was not over."

Oh Heaven!—I then felt the knife racking against the breastbone—scraping it! . . . and again began the scraping! The instrument this second time withdrawn, I concluded the operation was over— Oh no! Presently the terrible cutting was renewed—and even worse than ever, to separate the bottom, the foundation of this dreadful gland from the parts to which it adhered. Afterward, Baron Larrey was pale, nearly as me, his face streaked with blood and its expression depicting grief, apprehension, and almost horror.

The horrors endured by this remarkable woman mirrored the feelings of many thousands of other sufferers on whom this brutality was wreaked. A half century later, the discovery of anesthetic gases ended the terror of surgery performed without anesthesia.

REFERENCES
Burney, Fanny. *Selected Letters and Journals*. 1st ed. Oxford: Oxford University Press, 1986.

2

Ether and Nitrous Oxide: Blessed Relief

Pain humbles the proudest and subdues the strongest.
—Valentine Mott, surgeon, 1865

Introduction

Humphrey Davy was a son of a middle class family who were not overly concerned with the lad's education. But the boy was unusually bright, and he developed a love of chemistry. At an early age, when full of mischief and with a penchant for explosions, he created firecrackers, which he exploded unexpectedly. When his father died, the sixteen-year-old Davy began a course of remarkable self-education. During his apprenticeship with Bingham Borlase, an apothecary-surgeon in Penzance, Davy began writing poetry and reading history, theology, philosophy, and metaphysics.

Borlase allowed Davy to create a chemistry laboratory performing basic chemical research and to read Antoine Lavoisier's *Traité Élémentaire de Chimie* (Elementary Treatise on Chemistry) in French. Occasional explosions emanated from the laboratory, but fortunately no one was injured. While still an apprentice, Davy met Davies Gilbert, an Oxford graduate and well-to-do scientist, who was so impressed with Davy's brilliant intellect that he allowed Davy to use his well-stocked private library as well as introduced him to scientists, including Thomas Beddoes, who had settled in Bristol.

Beddoes was a learned scholar and a political radical. During his appointment at Oxford as a chemistry reader, he had followed closely the events in France, beginning with the Revolution and the Reign of Terror. He gained wide public attention with his preference for revolutionary ideals and his anti-British government pamphlets. Eventually, the British Home Office suggested that Oxford terminate Beddoes's employment, and he relocated to Bristol. In Bristol, Beddoes's pro-French political leanings were still widely known, and he was often rumored to be a secret seditionist. Some doubt was even raised about the quality of his scientific teachings, since he had often expressed his admiration for French scientific progress, including Lavoisier's contributions.

Yet Beddoes was able to make friends with several influential members of the Lunar Society of Birmingham, a selected group of intellectuals from many different fields who met monthly on the night of a full moon. Several wealthy men of the Lunar Society were taken by Beddoes's idea of building a research institute in Bristol, for which they provided generous financial support. Beddoes was a physician by training, and his friendship with these wealthy men came about because he had served as physician to their families.

In 1798, the Pneumatic Institute opened. Its stated aims were to study the therapeutic effects of the local "airs" and various gases in the treatment of disease. Although Beddoes had enlisted James Watt of steam engine fame to construct all of the needed apparatus, he also needed an assistant. Davies Gilbert recommended Davy, and Beddoes was so impressed with the young man's brilliance that he arranged for his release from his apprenticeship with Borlase. At the age of nineteen years, Humphrey Davy became the institute's superintendent of experiments.

Davy and Nitrous Oxide

Davy began studying the properties of the gases he isolated, and he was a compelling lecturer in summarizing his work and that of others. He soon became known as an invited speaker for Bristol society events. Davy had a long-time interest in nitrous oxide, dating back to 1775, when he had first experimented with the gas. As superintendent, Davy expanded his interest in the gas by studying its effects on himself, and he described one of his nitrous oxide "highs" as follows:

> The thrilling was very rapidly produced. The pleasurable sensation was at first local, and received in the lips, and about the cheeks.

It gradually, however, diffused itself over the whole body, and in the middle of the experiment, was as intense and pure as to absorb existence. At this moment and not before, I lost consciousness; it was however quickly restored, and I endeavored to make a bystander acquainted with the pleasure I experienced by laughing and stomping.

Davy offered this experience to such prominent citizens of the day as Samuel Taylor Coleridge, Robert Southey, and Peter Mark Roget (of thesaurus fame). It was during one such "experience" that Davy detected one of these guinea pigs giggling, and he immediately called the gas "laughing gas." In 1800, Davy published his meticulous observations in *Researches, Chemical and Philosophical: Chiefly Concerning Nitrous Oxide, or Dephlogisticated Nitrous Air, and Its Respiration*, a book that is regarded as a classic in the history of science.

Davy also found that a few whiffs of the gas relieved the pain of his seriously inflamed painful tooth, and he wrote the following prophetic words: "As nitrous oxide in its extensive operation appears capable of destroying physical pain, it may probably be used with advantage during surgical operations in which no great effusions of blood take place." Had Davy or any of the learned physicians of the day followed up on this speculation, the dawn of effective general anesthesia might have occurred a half century sooner. But no one did. Davy soon left the Pneumatic Institute and pursued a spectacular career in science. Lacking its young genius, the institute collapsed.

The Itinerant Showman

The giggling effects of inhaling nitrous oxide soon became the wares of traveling showmen as well as the centerpiece for private "laughing gas parties." At such affairs, guests inhaled a few whiffs of the gas, became giddy and silly, and fell down, laughing as if they were drunk. The antics produced by laughing gas soon became entertainment at shows for general public audiences. "Doctor Coult of New York, London and Calcutta," actually Samuel Colt, was one of the most colorful of the traveling nitrous oxide showmen. Colt went around the country carrying his nitrous oxide "laboratory" on a horse-drawn cart. Stopping on street corners, he would demonstrate the effects of the gas on himself and then invite spectators to try it for themselves, for a fee of course. Colt also sold bottles of the gas and encouraged the onlookers to try it at home. He used the money from his sales to develop and patent the revolver pistol that still bears his name.

Building on his success as an itinerant showman, Colt bought a part ownership of the Penny Museum in Cincinnati, where the entertainment consisted of the antics of people who were paid to stand on the stage and inhale laughing gas. But laughing gas didn't always produce laughter. At one performance, Colt hired six Native Americans to appear in what he thought would be a gas-inspired comedy. Upon inhaling the gas, the six promptly fell asleep on the stage without uttering a single giggle or whoop. Not to be taken back, Colt administered the gas to an obliging blacksmith who began chasing Colt around the stage. In his pursuit, the blacksmith tripped over and woke the Native Americans, who were surprised to find themselves on the floor.

Horace Wells

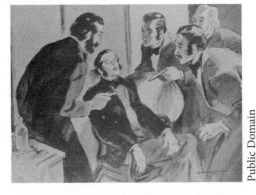

Public Domain

"Professor" Gardner Quincey Colton, a clever, partly trained physician, was a talented showman. He rented large halls in various cities to hold exhibitions of "Exhilarating or Laughing Gas," during which he featured volunteers from the audience who inhaled the gas on stage and amused the audience with their antics. At one such show in Hartford in December 1844, Horace Wells, a local dentist, volunteered to be a subject along with young James Cooley, who joined Wells on the stage.

Cooley first inhaled the gas and in his drowsiness, he fell against some furniture on the stage, which resulted in serious bruises on both legs. When he awoke, Cooley was surprised to see that his legs were bloodied. When questioned by Wells, Cooley remarked that he had felt no pain.

Dr. Wells immediately saw that the nitrous oxide could be used for the extraction of teeth without pain. In fact, at the very time, Wells himself had a painful molar that would require extraction. The next day, Colton delivered a bag of nitrous oxide to Wells's dental office, and Dr. John Riggs, an associate of Wells, painlessly extracted the tooth. Recovering from the anesthesia, Wells sat up in the chair and proclaimed, "Aha, a new era of tooth pulling!" He quickly learned the technique for making the gas, and he devised an apparatus for its administration that consisted of a bellows containing the gas and a tube inserted into the patient's mouth. Over the next few weeks, Wells and Riggs administered the gas to fifteen patients, all except two of whom underwent painless tooth extractions. It was the first definitive medical use of nitrous oxide, and it launched Wells's life-long quest for recognition as the discoverer of anesthesia.

Wells felt that his discovery had to be brought before the public. In February 1845, he traveled to Boston, where one of his former dental students, William Thomas Green Morton, had arranged a demonstration in the Massachusetts General Hospital, then as now one of the most prestigious institutions in the nation. Morton had persuaded the foremost surgeon in Boston, Dr. James Collins Warren, to allow Wells to demonstrate the effects of nitrous oxide on a volunteer patient who required a tooth extraction.

The surgical amphitheater was packed with skeptical physicians and medical students. When the patient was brought in, he suddenly became wary that he was the subject who was to receive an unknown gas delivered by a strange dentist, and he declined to proceed. It took several days before another student with a painful molar agreed to have it removed while receiving nitrous oxide.

Again the audience of physicians and medical students crowded into the operating amphitheater. Wells connected his apparatus, and the student promptly fell asleep. But just as Wells was beginning to extract the tooth, the student suddenly jerked his head back and cried out in pain. Roars of "humbug" rose over the amphitheater. The embarrassed Wells rushed from the amphitheater in disgrace and returned to Hartford on the next train. Later, Wells wrote that the patient had not been properly anesthetized, because the gas had been withdrawn too soon. This potentially pivotal moment in the history of anesthesia faded into historical insignificance except for Wells, who continued his search for public acclaim. But the flawed demonstration so depressed him that he sold his house in Hartford, sold his practice to Dr. Riggs, abandoned his wife and family, and set off to seek the recognition that he felt he had been so unjustly denied.

A second "exhilarating" gas was also gaining recognition. In those days, it was called sulfuric ether. Three centuries before nitrous oxide, Valerius Cordus, a twenty-five-year-old Prussian botanist, had prepared ether by adding sulfuric acid ("the sour oil of vitriol") to ethyl alcohol (the "strong biting wine") and produced what he called the "sweet vitriol," later named sulfuric ether and still later simply ether. Since ether was so simple to make, it became readily available to be used for pleasure or for drowsiness. By the mid-1800s, the destinies of ether and nitrous oxide were to become closely intertwined.

In 1841, four years before Wells was using nitrous oxide, another chapter in this unfolding drama in the history of anesthesia was occurring in the small Georgia town of Jefferson. Dr. Crawford Williamson Long, a recent graduate of the University of Pennsylvania Medical School, had launched his medical practice in his hometown of Jefferson. During his stay in Philadelphia, Long had joined other students in "ether frolics," and once he had settled in Jefferson, he hosted such affairs. The ether parties of that era would recognize their counterpart today as recreational drug-using affairs. At one of Long's parties, these fun-loving people asked Long to prepare some nitrous oxide for them to enjoy.

According to Long, "I informed them that I had no apparatus for preparing or administering the gas, but that I had a medicine (ether) that would produce equally exhilarating effects." He reassured them that he had inhaled it himself and considered it as safe as the nitrous oxide gas. "I gave it to all persons present. I noticed my friends, while etherized, received falls or blows, which I believe were sufficient to produce pain on any person not in a state of anesthesia. On questioning them, they all assured me that they did not feel the least pain from these accidents."

On March 30, 1842, Long excised one of two tumors from the neck of a Mr. James Venable under sulfuric ether anesthesia. Venable had repeatedly postponed the operation for fear of the pain, but Long promised him that the surgery would be painless. When the operation was over, Venable was incredulous until Long showed him the excised cyst. He had rendered his patient insensible using an ether-soaked towel. Some medical students and several other onlookers watched. On June 6, Long excised Mr. Venable's second tumor.

Other pain-free operations on different patients under ether anesthesia followed. Long recorded the operation in his office records: "James Venable; March 30, 1842, Ether and excising tumor, $2.00." But since he had waited seven years before publishing his experience, Long missed the opportunity to spread the word that would have otherwise made surgical anesthesia with ether a practical reality then and there. It happened four years later in October 1846 and not by Long, but by a Boston dentist named Morton.

William T. G. Morton

William Thomas Green Morton, the same man who eighteen months earlier had arranged Wells's disastrous demonstration of nitrous oxide, now claimed the spotlight. The twenty-seven-year-old Morton was practicing dentistry in Boston, and he wanted some way to relieve the pain of his patients who were being fitted with dental plates. Morton was a boarder in the home of Charles Jackson, a forty-two-year-old chemist, physician, and geologist. When he met Samuel Morse onboard a ship in 1832, Jackson allegedly explained to Morse the principles of the telegraph, or so he later recalled. Back in Boston, Jackson set up a private medical practice. But he gave it up in 1836 to establish a private laboratory dedicated to analytic chemistry. In 1844, Jackson demonstrated before several of his chemistry classes that inhalation of sulfuric ether causes loss of consciousness. Morton had been one of his students and had witnessed this demonstration.

Jackson suggested that Morton try ether, but the first few attempts were not satisfactory. Jackson then suggested that the uneven degree of purity of the ether that Morton had used was at fault, and he urged Morton to use a purer form of ether. Morton followed this advice and painlessly extracted a tooth from an etherized patient on September 30, 1846. Morton now knew that ether was the answer, but he kept it a secret. The next day, the following anonymous notice appeared in a Boston newspaper: "Last evening, as we were informed by a gentleman who witnessed the operation, an ulcerated tooth was extracted from the mouth of an individual without giving the slightest pain. He was put into a kind of sleep, by inhaling a preparation, the effects of which lasted about three quarters of a minute, just long enough to painlessly extract the tooth."

Morton was not only a bold dentist, but he was also a gifted publicist and a sly businessman. He promptly filed a patent application for his procedure for administering ether with the expectation of reaping large monetary rewards in the future. For the next three weeks, Morton used ether on several other patients, but the results were still not uniform. Morton faulted the apparatus he was using to administer the gas, and he designed a new device that provided a more even flow of ether.

Morton again approached Dr. James Collins Warren, seeking permission for a demonstration of ether anesthesia that he would administer. Warren agreed and set an early date. The time was short, so short that on the day of the operation, Morton arrived in the amphitheater fifteen minutes late, having been held up by last-minute changes in perfecting his apparatus. The amphitheater was once again filled with a medical audience, and remembering Wells's failed demonstration, was now even more skeptical. The patient was one Gilbert Abbott, who had a tumor on his left jaw. Morton promptly administered the ether, and Abbott fell fast asleep. Morton looked up at Dr. Warren and uttered these memorable words: "Sir, your patient is ready." The uneventful operation lasted twenty-four minutes. After regaining consciousness, Abbott told the audience that he had felt no pain whatsoever. Warren looked up at the audience and announced, "Gentlemen, this is no humbug."

In that single moment, Warren's powerful statement changed the course of surgical practice throughout the world. In the entire history of medicine, there is no discovery that can claim to have had such a momentous, immediate, and worldwide effect as those events that took place on October 16, 1846, in what is now known as the Ether Dome of the Massachusetts General Hospital. Oliver Wendell Holmes, professor of anatomy at Harvard, appreciated the wonder of anesthesia when he wrote in his typical flowery prose, "By this priceless gift to humanity, the fierce extremity of suffering has been steeped in the waters of forgetfulness, and the deepest furrow in the knotted brow of agony has been smoothed forever." He was right.

Morton's Next Moves

Morton's business instincts now took over. He intended to license the invention, no doubt for generous fees. He composed a letter to the hapless Wells relating the success of his demonstration and hinting that Wells might wish to become a licensee. The letter surely aggravated Wells's growing mental instability. Charles Jackson, Morton's landlord, sought to share in Morton's patent application. A bitter conflict arose between the two, and the patent was finally issued in both names, although later Jackson agreed to relinquish his stake in exchange for 10 percent of the royalties.

Jackson's claim was not the first time he had sought credit for what turned out to be incidental suggestions. Five years earlier, he had contested Samuel B. Morse's patent for the telegraph, and even earlier, he had claimed credit for the invention of guncotton following the announcement of C. F. Schönbein, the legitimate inventor. Jackson's eccentricities were to become even more obvious with age, and his animosity toward Morton would reach savage proportions as the news of Morton's fame spread.

Morton's patent was so stringent that his invention could not be used without his specific case-by-case permission. Reacting to this restriction, Dr. Warren invoked a moratorium on its use in the Massachusetts General Hospital. Morton relented, but he persisted in keeping his invention secret by using aromatic essences as a (poor) disguise of the ether. He admitted that the agent he used contained ether, but he denied that ether was the agent producing the anesthesia!

Drs. Henry Bigelow and Oliver Wendell Holmes were two of the most distinguished representatives of the hospital. Holmes suggested that the agent be named Letheon, apparently unaware that the term had been used by the ancients for the juice of the poppy. He also suggested the term *anesthesia* be adopted as well, although that term had also been used as far back as the Greeks.

Scarcely a month after Morton's initial triumph, Bigelow published a paper in the leading British medical journal, the *Lancet*, announcing the details of this new discovery to the world. Morton's secrecy was broken. News of the discovery spread quickly, and within months, it was hailed as the "greatest gift ever made to suffering humanity." At first, the English physicians were skeptical, and they dubbed ether the "Yankee dodge," but soon the Yankee dodge proved to be far superior to any other means of anesthesia and especially to mesmerism, which was then in vogue. Following an amputation with a patient anesthetized by ether, the great English surgeon James Liston remarked, "The Yankee dodge, gentlemen, beats mesmerism hollow."

Striving for Recognition

Jackson did not rest in his crusade for greater recognition as the discoverer of anesthesia. Through the good offices of a friend in Paris, he managed to have his case presented before the Académie des Sciences. The crux of his case was that he had masterminded the discovery of ether and that Morton had merely been his agent in carrying it out. Not to be outdone, Wells also traveled to Paris with a petition that he presented to the Société de Médecine de Paris. His case was weak since he had presented no confirming evidence to counter his dramatic failure.

These continental forays greatly upset Morton, who set about collecting sworn statements from witnesses to bolster his own case. Jackson came under a great cloud of suspicion at home in his unsuccessful attempt to persuade the American Academy of Sciences that he was the true discoverer of ether, even though his claim was backed by no less than the president of Harvard. Wells tried to bolster his case in Paris by later sending testimonials in support of his petition.

Meanwhile, Morton published an article laying out the details of his use of ether, and he sent a copy of the article to the Académie des Sciences in support of his claim. By this time, Wells's fragile psyche was fast approaching a breaking point. After leaving his family without any support, he moved to New York, where he tried to establish a dental practice offering nitrous oxide, ether, and chloroform anesthesia for dental extractions. Chloroform had only recently been introduced in the United States from Edinburgh, where it was used principally for anesthesia during childbirth.

Like ether and nitrous oxide, sniffing chloroform produced exhilaration, and Wells had become a chloroform addict. His new dental practice was summarily interrupted by his being jailed on the charge of throwing sulfuric acid at some ladies of the night. Wells, despairing and delusional and under the influence of chloroform, wrote a final letter to his wife, slashed the main artery in his thigh, and died in jail. A letter arrived a few days later stating that the Société de Médecine de Paris had anointed him as the discoverer of anesthesia!

After Wells's death, Paris erected a monument honoring him as the discoverer of anesthesia. Wells's hometown of Hartford also honored him with a statue. Certainly, Wells's use of nitrous oxide in dental surgery predated Morton's use of ether. But like Crawford Williamson Long, Wells did not publish his results, nor did he conduct any successful public demonstration of his finding. Wells may be credited as an innovator but not necessarily as the discoverer of anesthesia.

The Prizes

The Board of Trustees of the Massachusetts General Hospital prepared an official report that conferred upon Morton the mantle of discoverer of anesthesia and a monetary prize of $1,000. This would have seemed to settle the issue, except for a bill introduced in 1854 in the US Congress that promised a $100,000 prize to the discoverer of anesthesia, if one could be determined.

The first three claimants were the heirs of Morton, Jackson, and Wells. Crawford Williamson Long had finally published his results in 1849, under the imposing title "An Account of the First Use of Sulphuric Ether by Inhalation as an Anesthetic in Surgical Operations." He was out to establish his priority of discovery, citing his office records in 1842 as proof that he was the first to conduct surgery under ether anesthesia. Long became the fourth claimant. In the contest for the prize, Long asked Senator William Dawson of Georgia to present his case. Ironically, Senator Dawson chose Charles Jackson (of all people!) to investigate and report on Long's claim. Jackson traveled to Georgia, examined Long's records, and concluded that Long's claim was legitimate.

This conclusion almost certainly was another of Jackson's many attempts to discredit Morton. Jackson realized that his own claim was feeble, and he suggested that he and Long now file a joint claim, but Long refused. Long's hat had been thrown into the ring at a late stage of the debate, and this complicated matters to the point that the bill eventually died. The public was becoming weary of the controversy. Ether had triumphed, even though none of the claimants succeeded in establishing unequivocal priority. Oliver Wendell Holmes, perhaps thinking that Morton or Long had the best case, acknowledged that the controversy would never be settled, and he adroitly concluded that the honor of being the discoverer should go "to e(i)ther."

Their Futures

Morton, who should have gloried in the national esteem that was being heaped upon him, spent the remainder of his life in the courts arguing over many infringements of his patent rights. These litigations proved to be an increasingly heavy mental burden that was made worse by a censure from the American Medical Association and by Jackson's continuing bitter public opposition. It was too much for Morton to bear. On a carriage ride through Central Park in New York with his wife, Morton suddenly bolted the carriage and plunged his head into a nearby lake. Moments later, he again jumped out the carriage, leaped over a stone fence, and fell unconscious. He was taken to St. Luke's Hospital, where he died a few hours later. Morton was buried in the Mount Auburn Cemetery near Boston.

Charles Jackson fared no better. He experimented with ether on cases of insanity at the McLean Insane Asylum and published his results in a *Manual of Etherization*. It received little favorable attention. His mental condition grew progressively unstable to the point that he was admitted to McLean as a patient. He was completely insane and spent the last seven years of his life in the institution. There is no record of whether ether was prescribed for him or for any other of the McLean patients.

Crawford Williamson Long was the only one of the four to have escaped the mental "curse" that had brought down the other three. He continued his practice, and during the Civil War he cared for a great many wounded Confederate soldiers. During one period of fierce fighting, the Confederate Army was forced to make a rapid retreat from the advancing Union troops. Long scurried to the rear carrying a glass jar that contained "my proofs of the discovery of ether anesthesia."

Long died in 1878 at the age of sixty-two. He was in the process of delivering a baby from an etherized mother, and as he handed the newborn infant to an attendant, he suffered a fatal stroke. By an act of the Georgia General Assembly in 1920, Long County was named after Long. He is also honored, as is Morton, in the National Hall of Fame in Washington, D.C.

The Monument to Ether

In 1868 Thomas Lee, a Bostonian, sponsored a contest to build a monument to commemorate the real discovery of ether anesthesia that would stand in the Boston Garden. When sculptor John Quincy Adams Ward set about creating it, he recognized that there was a major dispute over who was the discoverer. Ward skirted the controversy by depicting a medical doctor in a medieval Moorish-Spanish robe and turban, representing a Good Samaritan, who is holding the drooping body of an almost naked man on his left knee atop the granite and red marble monument.

The Moorish doctor holds a cloth in his left hand, suggesting the use of ether that would be developed in centuries to come.

The use of a Moorish doctor was almost certainly intentional, thereby avoiding choosing any sides in the debate that was raging at the time over who should receive credit for the first use of ether as an anesthetic. Rather than bearing any name, an inscription on one side proclaims, from Revelation 21:4, "Neither shall there be any more pain." An inscription on another side reads, "To commemorate the discovery that the inhaling of ether causes insensibility to pain—first proved to the world at the Mass. General Hospital in Boston—October AD MDCCCXLVI."

Comment

Chance played decisive roles in the discovery of both nitrous oxide and ether as anesthetics. In 1841, Crawford Williamson Long, starting his medical practice as a young doctor, threw a party at which his guests enjoyed the exhilarating effects of ether. One guest under ether's influence confessed to feeling no pain after injuring his leg while falling, and Long immediately saw ether's application to surgery and used it for an operation to painlessly remove two cysts from the neck of James Venable in March 1842.

Horace Wells had a similar chance experience and arrived at the same conclusion. By some good fortune, he happened to attend Professor Gardner Quincey Colton's laughing gas show in Harford in late 1844. James Cooley, a co-participant with Wells as volunteers, breathed the gas on the stage, fell against some furniture, and seriously injured his leg. Waking from his nitrous oxide sleep, he confessed to feeling no pain. For Wells, it was an accidental discovery meeting a prepared mind in his perfecting painless dentistry just as Long had with ether. Wells immediately had the best possible proof of painless tooth extraction by having one of his own molars extracted while he was under the influence of the anesthetic.

It's safe to assume that before Long and before Wells, many onlookers at laughing gas shows and ether frolics had seen episodes that were the same as those that Wells and Long had observed. Both men saw what others had seen, but their responses were different from the others because their minds were more receptive by their past training and current experience to be attuned to new ideas for relieving pain. Winston Churchill once said, "Once in a while, one will stumble upon the truth, but most of us manage to pick ourselves up and hurry along as if nothing had happened." Neither Long nor Wells were among those who hurried along.

REFERENCES

Erving, H. W. "The Discoverer of Anesthesia: Dr. Horace Wells of Hartford." *Yale J Biol. Med.* 5 (1933), 421–430.

Kenyon, T. K. "Science and Celebrity: Humphrey Davy's Rising Star." *Chem. Herat. Mag.* Winter 26 (2008–2009), 1–3.

Long, C. W. "An account of the first use of Sulphuric Ether by Inhalation as an Anesthetic in Surgical Operations." *South Med Surg J* 5 (1849), 705–713.

Papper, E. M. *Romance, Poetry and Surgical Sleep: Literature Influences Medicine.* Westport, CT: Greenwood Press, 1995.

Robinson, S. *Victory over Pain: The History of Anesthesia.* New York: Schuman, 1946.

Stansfield, D. A., and R. G. Stansfield. "Dr. Thomas Beddoes, and James Watt: Preparatory work 1794- for Bristol Pneumatic Institute." *Med Hist.* 30 (1986), 276–302.

Wells, H. "The history of the discovery, of the application of nitrous oxide, ether, and other vapors, to surgical operations." Hartford, CT: Gaylord Wells, 1847.

3

Chloroform: The Clergy Objects

It is a great crime to leave a woman alone in her agony and deny her relief.

—*Grantly Dick-Read, MD*

Introduction

The rustic hamlet of Sackets Harbor on Lake Ontario may seem an unlikely site for the discovery of chloroform. It came about because of a thirty-five-year-old physician, Samuel Guthrie, who had chosen to set up his medical practice in the village in 1817. But Guthrie soon found that rural medicine had much less appeal for him than his much deeper fascination with chemistry. Once settled, Guthrie set up a chemical laboratory in an outbuilding on his property, and he went to work, developing an excellent brand of home-distilled alcohol as well as a "Percussion Pill" that was used to fire weapons. Guthrie began a correspondence detailing his chemical researches with the famous Professor Benjamin Silliman at Yale, the author of *Silliman's Elements of Chemistry*.

Through the letters, Silliman recognized Guthrie as a gifted chemist. Silliman's textbook, the standard reference of its time, described chloroform, or "chloric ether" as it was called, as "heavy oil with a sweetish, aromatic and agreeable taste. Taken internally, it is stimulating and reviving, although its medical powers have not yet been ascertained." Inspired by this passage, in 1830 Guthrie devised and published a "New mode of preparation of a spirituous solution of Chloric Ether." His process was ingenious and inexpensive.

He sent the details to Silliman, who published it in 1831 in his journal. Guthrie's chloric ether, dubbed "Guthrie's sweet whiskey" by the Sackets Harbor locals, had intoxicating effects that rivaled those of alcohol. Silliman thought that chloric ether ought to be studied for its clinical effects, and he went so far as to distribute samples to some medical colleagues in New Haven. They expressed no interest in it whatsoever. The first observation of the anesthetic properties of chloric ether occurred when Guthrie's eight-year-old daughter accidentally tasted the sweet liquid in a vat in the laboratory. She quickly fell asleep on the floor. She soon recovered, but even though Guthrie was a physician, he failed to recognize the medical significance of his daughter's spell.

James Young Simpson

James Young Simpson, a Scot born in 1811, was to bring chloric ether, renamed chloroform, into the practice of medicine. In 1832, he completed his medical studies at Edinburgh University, a leading medical school at the time. The school was innovative in establishing a chair of midwifery, and

Public domain

its first occupant was a Dr. James Hamilton. Hamilton was an innovator who managed to earn the scorn of his peers by advocating the presence of men in the birthing room, an idea that was unheard of at the time. It was believed to be indelicate, even immoral. To underscore their point, in 1591 the Calvinists convicted a distinguished noblewoman, Lady Eufame Macalyene, for the crime of seeking pain relief from Agnes Sampson, her midwife, during her labor with twin sons. Sampson reported the request to the religious authorities, who found that she had violated the doctrine of the primeval curse on woman. She was burned alive on Castle Hill in Edinburgh as punishment.

Simpson became one of Hamilton's converts, even though as a student, he had slept through most of Hamilton's late afternoon lectures. Upon Dr. Hamilton's death, the twenty-eight-year-old Simpson applied for his now-vacant chair. After a bitter dispute, Simpson squeaked by with a single vote. Thanks to his winning personality and his skill, Simpson's obstetrical practice grew rapidly, and he attended many prominent women in Edinburgh, London, and on the Continent. In 1846, he heard of the use of ether in London, and he thought that its application to obstetrics was "a glorious thought." This idea of relieving the pain of childbirth was revolutionary to many religions that had totally rejected the idea on the grounds that it was God's punishment and not to be interfered with. Simpson's opponents bolstered their case by quoting Genesis 3:16: "The Lord God said to the woman, I will greatly multiply thy sorrow and thy conception; in sorrow, thou shalt bring forth children." But Simpson countered with scripture of his own from Genesis 2:21: "And the Lord God caused a deep sleep to fall upon Adam, and he slept; and He took one of his ribs, and closed up the flesh instead thereof."

In 1896, Andrew Dickinson White, a cofounder of Cornell University, published his *History of the Warfare of Science with Theology in Christendom*, in which he claimed that "from pulpit after pulpit, Simpson's use of chloroform was denounced as impious and contrary to Holy Writ; many texts were cited, the general declaration being that to use chloroform was to avoid one part of the primeval curse on woman." Simpson wrote pamphlet after pamphlet to defend the blessing of chloroform, which he had brought into use. Simpson argued that God was an anesthetist, since he caused Adam to sleep while He removed his rib! He chided the churchmen by noting that some of their past brethren had spoken strongly against eyeglasses and telescopes as "offspring of men's wicked minds." Now, however, these were fully accepted. Simpson was not to be dissuaded by his clerical opponents, and he promptly incorporated ether into his obstetric practice and soon became its enthusiastic advocate.

But the use of ether was still at an early stage, and there were some serious drawbacks. It irritated the lungs and often caused vomiting in patients when they awoke. Of greater importance, it was extremely flammable and explosive. For short procedures like tooth extraction, where a handkerchief merely moistened with ether would be sufficient, the risks were low. But when it was used for a longer procedure, such as childbirth, the birthing room was soon filled with ether fumes, sufficient to intoxicate the doctor and his attendants.

Not only that, but chance exposure to the flame of the many candles and gas lamps that provided illumination could easily result in a devastating explosion and fire. Finally, there was the very practical objection that the heavy glass containers of ether were quite a burden to carry up the stairs of the overcrowded tenements of Edinburgh, particularly since a large number of such bottles would be needed if the labor were prolonged.

Simpson's Experiments with Chloroform

Simpson began searching for an alternative to ether by trying a number of volatile chemicals on himself and on two of his associates, Drs. Mathew Duncan and George Keith, but without any promising results. In the fall of 1847, as Simpson happened to be chatting with David Waldie, an apothecary, he learned that Waldie had been dispensing chloroform to local physicians for use as an inhalant for various respiratory disorders. Waldie also mentioned in passing that several patients fell asleep after receiving chloroform inhalations. Simpson immediately saw that chloroform would be an excellent replacement for ether, and he was eager to test it.

Chloroform was a simple compound that, thanks to Samuel Guthrie's work, could be easily made. It was less volatile than ether, and it had a pleasant, sweet odor. Simpson set out to test the anesthetic properties of chloroform, and he invited his two colleagues to participate in an experiment. The three men sat around the Simpson family dining table, and Simpson poured the liquid into three tumblers. He and his two friends inhaled the fumes from the tumblers as deeply as they could until they lost consciousness.

Simpson's wife, Jessie, stood by to record the exact time at which each of three became unconscious. After a few breaths, Simpson fell into a drunken state, and his two colleagues became exhilarated and then collapsed. Simpson and Keith slid off their chairs onto the floor asleep while Duncan bent over the table fast asleep. Jessie was alarmed, and she seriously considered whether she should try to wake the sleeping trio. But Simpson had left strict instructions that they were not to be disturbed. When he awoke, he saw that Duncan was peacefully bent over the table snoring loudly, while Keith was violently kicking the underside of the table. Shortly thereafter, Simpson persuaded his niece, Miss Petrie, to try it. After inhaling the vapors, she promptly fell asleep singing the words, "I am an angel!"

Simpson knew that chloroform was what he was been looking for. He wrote to the Medico-Chirurgical Society pointing out chloroform's advantages over ether. It was more potent, easier to administer, provided quicker anesthesia, was not flammable, and had minimal aftereffects. He wrote of his first use of chloroform on an obstetrical patient: "I placed her under the influence of Chloroform by moistening with a half a tea-spoon of the liquid on a pocket-handkerchief that was rolled up with the broad or open end of the funnel placed over her mouth and nostrils. The child was expelled in about twenty minutes. When she awoke, she observed to me that she had enjoyed a very comfortable sleep."

The child was a healthy baby girl whose father was a physician and a friend of Simpson's. The parents were so elated at the painless birth that they named their daughter Anesthesia. Some years later, Simpson received a photograph of the adult Anesthesia, which he proudly displayed above his desk. Many other women followed this first chloroform patient. Patients receiving chloroform weren't groggy and had no nausea or vomiting when they awoke. It seemed that painless childbirth had arrived.

Yet for a time, chloroform was not widely accepted for easing the pains of labor. This had little to do with its possible adverse influence on uterine contractions or on the health of the newborn infant. One main objection from women stemmed from a few instances in which chloroform anesthesia produced sexual passions. Despite such sporadic criticism, chloroform's fame grew. Physicians from all over Britain journeyed to Edinburgh to witness the use of chloroform during childbirth. At one demonstration, the bottle of chloroform fell to the floor and drenched the carpet. Simpson quickly cut a piece of the rug and used it to administer the anesthesia. As Simpson's reputation grew, he received honors and recognition both in Britain and elsewhere.

But there were higher risks associated with chloroform than with ether, and its administration required greater physician skill. There were early reports of fatalities due to chloroform, beginning with a fifteen-year-old girl in 1848. Skill, care, and experience were required to differentiate between an effective dose—i.e., enough to make patient anesthetized during surgery—and one that paralyzed the lungs, causing death. Still, use of chloroform spread quickly, and in 1853 it was famously administered to Britain's Queen Victoria during childbirth.

During the Mexican-American War (1846–1848), doctors began using choroform as an anesthetic on the battlefield; by 1849; the US Army officially began using it. Chloroform received a boost for use in military medicine as a result of its successes achieved during the Crimean War (1853–1856). Although many army doctors had prior experience with ether, by the time of the US Civil War (1861–1865), chloroform rapidly became the preferred anesthetic for battlefield injuries. It was easier to administer, its effects were faster than ether's, and the postoperative effects were less severe than with ether.

Simpson the Man

Queen Victoria had a special affection for Scotland and specifically for James Young Simpson, whom she appointed as physician to the queen in Scotland. In 1866, Her Royal Highness knighted Simpson in gratitude for his discovery of chloroform anesthesia. Thereafter he adopted the motto of *Victo dolore* ("Victory over Pain") for his coat of arms. Having the praise of the sovereign instantly made Simpson a national hero, and it quelled all of the ideas that anesthesia was an invention of the devil or that a man could not be present at the birth of a baby.

Simpson was an inspiring and vigorous personality. He was beloved by his patients. His sympathetic manner appealed to all he met. He was always ready to care for the poor and gave much of his time to them. Simpson was ever the champion of chloroform. When the *Encyclopedia Britannica* asked him to write the section on anesthesia, he wrote it to appear under the C's for chloroform and not under the A's. In his article, Simpson made light of the American discoveries of the use of ether and nitrous oxide. In 1847, when the Lord Provost of Edinburgh awarded Simpson the Freedom of the City, he introduced him as the discoverer of not only of chloroform but also of all anesthesia. Simpson did not correct him. Along with many others, he truly believed that he deserved that accolade.

Simpson was a devoutly religious man. At every breakfast, the family, each member with a Bible, gathered around the table for prayers. After the tragic death of Jamie, his fifteen-year-old son, Simpson had a moving encounter with Jesus that further fortified his faith. His beliefs supported him through a severe and painful illness during his last months. He died at the age of fifty-nine in his house on Queen Street. Edinburgh declared a day of mourning for its national hero. Flags flew at half-mast, and the stock exchange, the banks, the university, and all of the shops and businesses closed. The solemn bells of St. Giles pealed over the city as 100,000 sobbing Scots crowded onto the streets.

Although Simpson could have been buried at Westminster Abbey, his family preferred that he be laid to rest in Warriston Cemetery. A prominent statue of Simpson now still stands on Princes Street in Edinburgh and his bust resides in the abbey. Upon Simpson's death, a colleague wrote, "Simpson adopted obstetrics when it was the lowest and most ignoble of our medical arts, and he left it a science numbering amongst its professors, many of the most distinguished of our modern physicians."

John Snow and Queen Victoria

In London a physician named John Snow, who was nearly the same age as Simpson, began some scientific studies of anesthesia. Snow had developed a scientific mind in evaluating what he observed. In 1854, a cholera epidemic struck the London area of Broad Street, claiming some several hundred lives within a few days. Snow visited Broad Street and learned that the almshouse and the brewery were the only places that had been spared. He discovered that neither place had used any water from the Broad Street public water pump. He concluded that the source of cholera was the contaminated pump water that was being drawn from the Thames. He persuaded the Parish Board of Guardians to have the handle of the pump, which was used by all of the residents of Broad Street, removed. Three days later, the cholera epidemic of Broad Street came to a halt.

Even though ether had been introduced some years before in London, its initial popularity had fallen dramatically because of some side effects and because the techniques of its administration were faulty. Snow began an "ether practice," and teaming up with an influential surgeon, he rapidly gained recognition as the man who could correctly, effectively, and safely administer ether anesthesia. When Snow heard the enthusiastic reports of chloroform from Edinburgh, he tried it at once and was so pleased with it that he discarded ether in favor of chloroform. Snow then set out to makes meticulous scientific measurements of the effects of chloroform anesthesia in order to develop precise and reproducible methods for its administration.

In 1853, an invitation came for Snow from Prince Albert, informing Snow that Queen Victoria was to have another baby, and she wished to receive chloroform to relieve the pains of her delivery. Snow administered chloroform to the queen during the birth of Prince Leopold in April 1853. The delivery took fifty-three minutes, after which Her Majesty remarked, "The effect was soothing, quieting and delightful beyond measure." The queen was so pleased that Snow was again summoned two years later for the birth of Princess Beatrice. Because of his great notoriety in attending the sovereign, Snow was deluged with questions about the details of the queen's deliveries. He answered them all with the same good-natured remark: "Her Majesty was a model patient."

One lady to whom he was about to administer chloroform stubbornly declared that she would not inhale chloroform unless she was a given a word-by-word account of the queen's delivery. Snow replied, "Her Majesty asked no questions until she had breathed very much longer than you have; and if you will only go on in royal imitation, I will tell you everything." The patient could not but follow the regal example. In a few seconds she was asleep, and by the time she awakened, Snow had already left the house.

John Snow's death at the age of forty-five was not totally unexpected, since he had been suffering from both kidney disease and tuberculosis for several years. Some of his colleagues believe that his premature death was related to his zeal for frequent self-experimentation with various anesthetic agents, some of which were probably dangerous. Despite his generally poor health and obliviousness to the honors that he had received, Snow had dedicated his life to studying and administering anesthesia. He attended over 450 patients during his lifetime.

Snow was a solitary man, unmarried and without family, who preferred to devote his time to his studies rather than take time for a holiday. Snow suffered a slight stroke in early June 1858, while finishing a book on anesthesia. He was able to complete the work before he suffered a second fatal stroke only a few days later. After Snow's death, Benjamin Richardson (later Sir Benjamin) prepared Snow's manuscript for publication with the title *On Chloroform and Other Anesthetics*. This book is now regarded as the first textbook of anesthesiology, and its author is deemed to be the first scientific anesthesiologist.

Snow was buried in London in the Brompton Cemetery, located a short distance from the site of his famous study of cholera. Soon after his death, his friends erected a monument, which itself has a unique history. The initial inscription contained an error for his date of birth, which was listed as March 15, 1818, rather than March 15, 1813. The deteriorated monument was restored in 1895 and again in 1938, but each time the error went undetected. During World War II, a bomb damaged the monument during an air raid. After the war it was once more restored; this time Snow's correct birth date was inscribed.

Chloroform Fatalities

As the use of chloroform grew, so did the number of fatalities of patients on whom it was used. Before his death, Snow had collected fifty such instances from the United States, Britain, and the Continent over a period of seven years. Many more were to follow. The first use of chloroform in Germany was an anesthetic for a bear in the zoological gardens of Berlin. The old bear was hopelessly blind due to cataracts. The leading clinician of the day, Johann Schönlein, used his influence with the king to have the cataracts removed under chloroform anesthesia. The operation was successful but the bear never woke up. When the sculptor Wilhelm Wolff heard of this affair, he created a bronze group of various animals in which the large bear bore a striking resemblance to Schönlein, and an ape holding a chloroform bottle in his hand was an easily recognized portrayal of Hans Jüngken, the surgeon who had performed the operation.

The statue so perplexed King Frederick William IV that he authorized a prize to be given for the best explanation of it in verse. A young poet submitted the winning poem:

A dead man is the bear, I fear,
The reason, well, this chloroform here.
A consultation of medical men
Treated too heavily the animal then.

The foxlet grins, little bear will lament,
The wolf erects his monument.

Many of the chloroform deaths tragically involved minor surgical procedures. The fatalities most often occurred during the first stage of anesthesia and were probably due to overdosing. Chloroform has a narrow margin of safety and readily depresses the heart, blood vessels, liver, and kidneys. When the number of chloroform-related fatalities grew to alarming proportions, it fell completely out of favor. Fatal heart rhythms produced by chloroform were the chief culprits of the anesthetic deaths.

Comment

Sir James Young Simpson deserves an honored place in the history of obstetrics. He fought against what he considered to be erroneous beliefs from Scripture that women in labor should suffer and that men should never be allowed to attend women during childbirth. Considering his earlier and largely unsatisfactory experience in using ether for successfully easing the pains of labor, Simpson's fertile mind immediately jumped on the casual remarks of David Waldie and concluded that chloroform might be a better anesthetic. His experiments on himself and with two of his friends were vivid proof that chloroform was in fact a laudable new anesthetic. He became its champion, and he passed his legacy to John Snow, a remarkable scientific anesthesiologist. Although chloroform had a relatively short history as an anesthetic, Simpson's life and work served as a groundbreaker in obstetrics.

REFERENCES

Cranefield, P. F. "J. Y. Simpson's Early Articles on Chloroform." *Bull NY Acad Med.* 62 (1986), 903–909.

Hemple, S. *The Medical Detective: John Snow and the Mystery of Cholera*. London: Granta Books, 2006.

McCrea, M. *Simpson, Chloroform and the Spirit of Progress: A Biography of Sir James Y. Simpson*. Edinburgh: John Donald Publishers, 2010.

Simpson, J. Y. "An Account of a New Anesthetic Agent, Substitute for Sulphuric Ether in Surgery and Midwifery." *Lancet* 2 (1847), 549–550.

Vinten-Johanson, P. et al.: *Cholera, Chloroform, and the Science of Medicine*. New York: Oxford, 2003.

Cocaine Anesthesia: Freud's Favorite

"Coca Koller"
—*Sigmund Freud, in an inscription of a copy of* Über Coca

Introduction

I n 1858, Albert Niemann, a student of the distinguished chemist Professor Friedrich Wöhler at the University of Göttingen, successfully isolated and crystallized cocaine from the leaves of the coca tree. Niemann tasted the cocaine crystals and found that his tongue became numb and that he could not distinguish between hot and cold. But Niemann saw no medical significance to this finding, nor did any of the many others who confirmed his observation. In 1880, Vasili von Anrep, a distinguished pharmacologist, confirmed the numbing effects of oral cocaine and also discovered that it dilated the pupil of the eye. But he missed discovering its anesthetic effect on the eye.

He dismissed cocaine as having no medical uses. Indeed, the British Medical Commission thought cocaine was "a poor substitute for caffeine." Even so, Merck and Company began to produce cocaine in 1884 and distributed it to doctors for test purposes, among them a young Viennese physician named Sigmund Freud.

Carl Koller had graduated from medical school at the same time as his good friend Sigmund Freud. During his medical school days, Koller did some original experimental work in embryology under Professor Salomon Stricker. After graduation, Koller continued his good relations with Stricker, including having access to Stricker's laboratory. From the beginning of his training in ophthalmology, Koller realized that there was an urgent need for a means of anesthetizing the eye for surgery.

At the time, the usual practice was to perform the procedure without any anesthesia, an extremely painful operation that was made worse by the patient shrieking and thrashing about and being restrained by men using gags or a wooden spoon. Stitches were often torn out, and the results were disastrous for both patient and surgeon. General anesthesia was not satisfactory because of the postoperative restlessness and episodes of vomiting that disrupted the surgical repair, which for proper healing required the patient to lie still for long periods of time. Koller continued his search for the ideal local anesthetic and spent a year studying a variety of possible agents on animals. None showed any anesthetic properties.

Freud and Cocaine

Following medical school, Freud was deeply impressed with the tales of the Peruvian explorers that the coca leaves produced dramatic effects. He decided to take time from his duties as a young physician to study and publish *Über Coca*, a compilation of much that was known about cocaine. Thanks no doubt to his self-experimental use, Freud was quite enthusiastic about its properties.

Library of Congress

It was Freud who brought cocaine into medicine, albeit quite indirectly. Ernst von Fleischl, a close friend of Freud and a physician, had a severely injured, extremely painful stump of a left thumb. He started taking morphine for the pain, but the escalating doses needed for control of the pain reached dangerously high levels. An alarmed Fleischl sought help from Freud. Upset by his friend's desperate state, Freud substituted cocaine for morphine, believing that it would quell Fleischl's craving. This prescription worked for a while, but as larger and larger doses were required, severe chronic cocaine intoxication ensued and lasted over seven agonizing years before his death. Cocaine addiction had simply substituted for morphine addiction. This disastrous event marked the first medical use of cocaine.

Freud was especially impressed with the stories of the Incas' ability to perform incredible feats of stamina without food or sleep, particularly since he had already experienced some of these effects on himself. At the time, the notion that cocaine could impart high levels of energy had led some chemists to try to isolate the responsible agent. None succeeded. Coca was thereafter "authoritatively" considered as simply being a mild stimulant like tea or coffee, particularly since the chemical structure of caffeine was quite close to that of cocaine. Its alleged properties voiced by travelers to Peru were deemed to be legendary and fictitious, and the travelers themselves were often dismissed as ignorant or affected with superstitious imaginings.

But Freud thought otherwise, and he first approached Leonard Köenigstein, a house physician, with a proposal to test this energy-enhancing ability in some physiological experiments on "muscular strength, fatigue and the like." Köenigstein refused the offer. Freud then, recalling Carl Koller's access to Professor Stricker's laboratory facilities, made the same offer to Koller, also a house physician. Koller immediately agreed. Koller also had a particular interest in cocaine that dated back to his studies of pharmacology at the university. Koller recruited another colleague, a Dr. Gaertner, to join him in the study.

Before the experiments got under way, Freud departed for Hamburg on a long-awaited visit to his fiancée, Martha Bernays, leaving the two researchers on their own. Koller's daughter Hortense saved Koller's journal in which he recorded his first glimmerings of his discovery: "A colleague of mine partook of some [cocaine] with me from the point of his penknife." He remarked: "How that numbs the tongue. (I responded) Yes, that has been noticed by everyone, and in a moment, it flashed upon me that I was carrying in my pocket the local anesthetic [for the eye] for which I had been searching."

Koller first applied a few drops of a 2 percent cocaine solution in the eye of a guinea pig. At first, the animal winked, but within a half a minute, it opened its eye, revealing a staring expression. After a minute or so, Koller touched the cornea with a pin but saw no response. The eyeball remained entirely immobile. He then scratched and pricked the animal's cornea with a needle, irritated it with an electric current, and cauterized it with silver nitrate, but none of these insults provoked any discernible reaction.

Koller and Gaertner decided to test the solution on each other's eyes. Gaertner described the experiment: "We trickled the solution under the upraised lids of each other's eye. Then we put a mirror before us, took a pin in hand, and tried to touch the cornea with its point. Almost simultaneously, we could joyously assure ourselves by saying, 'I can't feel a thing.' We could make a dent in the cornea without the slightest awareness. I rejoice that I was the first to congratulate Dr. Koller as a benefactor to mankind."

Success and Recognition

Koller instantly knew that cocaine anesthesia represented a great advance in eye surgery. He persuaded a friendly eye surgeon to try cocaine anesthesia for the removal of a cataract. With cocaine anesthesia, the operation proved to be completely painless and uneventful. Upon his return from Hamburg, Freud was astonished and delighted with the results that Koller had obtained. On September 15, 1884, the German Ophthalmologic Society was meeting in Heidelberg. Koller was eager to present his results to this distinguished audience, but as a junior physician, he could not afford the trip. Koller had prepared a two-page handwritten paper and had asked a senior colleague, Dr. Joseph Brettauer, to read it at the meeting.

A day before the large general session was scheduled, Brettauer convened a small group of leading European and American ophthalmologists, and he presented the details of Koller's discovery, much to their astonishment and delight. Brettauer arranged for a patient from the Heidelberg Eye Clinic to be brought before the large audience the next day for a demonstration of cocaine anesthesia. The frightened patient was brought in and seated in a chair, and a few drops of a cocaine solution were dripped into one of his eyes. After a few minutes, Brettauer probed the eyeball, inserted a speculum to separate the eyelids, and grasped the eyeball with a forceps. The patient felt absolutely nothing!

Dr. Henry D. Noyes, the president of the American Ophthalmological Society, who attended the demonstration, was so impressed that he hurriedly sent back a report to the United States, which was quickly hailed in nearly every pharmaceutical and medical journal there. Of note was the response of the great Philadelphia surgeon Professor David H. Agnew, who wrote in the *Medical Record* on October 18, 1884, "We have to-day used the agent (cocaine) in our clinic at the College of Physicians and Surgeons with most astonishing and satisfactory results. If further use should prove to be equally satisfactory, we will be in possession of an agent for the prevention of suffering in ophthalmic operations of inestimable value." Just as William Thomas Green Morton had instantly revolutionized the whole field of general surgery with his use of ether anesthesia, so Carl Koller transformed the practice of eye surgery in a single moment with cocaine.

A month later, Koller presented his discovery to the Vienna Medical Society. What should have been Koller's crowning moment was marred when Leonard Köenigstein, the man who had refused Freud's offer to conduct the physiological experiments on cocaine, rose to challenge Koller by alleging that he and not Koller had discovered cocaine anesthesia for the eye during his studies of eye disease that had been originally suggested by Freud. He shouted that he was the one who deserved all of the credit.

Freud knew that Köenigstein had not discovered the anesthetic properties of cocaine, and he and Dr. Julius Wagner-Jouregg, another influential physician and future Nobel Prize laureate, convincingly persuaded Köenigstein to retract his false claim. Köenigstein backed down and agreed to write a letter to a medical journal confirming Koller's role in detail and disclaiming any credit for him. Freud rejected any small credit of the discovery for himself, and he wrote, "Koller is rightly regarded as the discoverer of local anesthesia by cocaine, which has become so important in eye surgery."

Koller's claim was secure, and his fame grew rapidly. Reports of cocaine's effectiveness as an anesthetic for operations on the eye as well as on the nose, throat, and teeth spread rapidly throughout all of Europe and the United States. "Kollerism," "Kollerization," and "Kollerize" soon became part of medical jargon to describe cocaine anesthesia. Freud inscribed a copy of his new edition of *Über Coca* to "*Seine liege Freund, Coca Koller*" (To my dear friend, Coca Koller). The nickname stuck.

Koller's Future

While cocaine's popularity increased, Koller's did not. Koller was certainly famous, but being Jewish and working in the hotbed of anti-Semitism of Vienna, he encountered widespread hostility. While serving a tour of duty as a junior physician in the hospital, Koller admitted a patient with an extremely tight dressing. Koller grabbed a pair of scissors and quickly released the bandage. Freiderich Zinner, a fellow physician, violently disagreed with Koller's action and called Koller "an impudent Jew." Koller immediately struck a resounding blow to Zinner's face, to which Zinner's rejoinder was a challenge to a duel. Even though Koller was no swordsman, he accepted the challenge. The two duelists agreed to use thin, lightweight, razor-sharp foils.

During the third round of the duel, Koller inflicted two large gashes on Zinner's forehead that required hospitalization. The duel added another hurdle to Koller's wish to pursue his career in Vienna. Both of Vienna's prominent eye clinics rejected Koller, who was regarded as a difficult and tempestuous young Jew. Deeply hurt, he moved to Utrecht for further training, and in 1888, he came to the United States and established a successful ophthalmologic practice at Mount Sinai Hospital in New York City. He abandoned cocaine research completely and devoted himself entirely to his busy practice.

Throughout his life in the United States, Koller impressed everyone with his competence as a skilled ophthalmologist. The Medical Association of Vienna honored him in 1930, and many honors, both in Europe and in the

United States, followed. He was proposed several times for the Nobel Prize in Physiology or Medicine. Certainly, his discovery of cocaine as a local anesthetic in ophthalmology was undoubtedly worthy of this prize, but the Nobel Committee felt that his discovery had been published too far in the past, so that according to the statutes of the Nobel Prize, this distinction could not be granted because it was deemed to be too old. Koller died in New York in 1944 at the age of eighty-seven.

Comment

Once more we observe a situation where a chance set of circumstances led to a major discovery. Köenigstein's refusal of Freud's offer to participate some physiological trials, and Freud's departure to visit Martha Bernays, set the stage for Koller and Gaertner to begin their experiments. Many before Koller had sensed the numbness of mouth when cocaine was place on the tongue, but only Koller's fertile mind grasped its significance as an agent that could produce numbness of the eye. The revolution of eye surgery was borne at that instant.

REFERENCES

Hall, M. "Coca Koller: The beginning of local anesthesia." *Anes Prog* 19 (1972), 65–67.

Reis, A. Jr. "Sigmund Freud (1856–1939) and Karl Köller (1857–1944) and the discovery of local anesthesia." *Rev. Bras. Anesthesia* 59 (2009), 244–257.

Part II
Implants

The Artificial Lens: A Legacy of Wartime

In his fight for sight, Ridley changed the world.

—*David J. Apple, MD*

Gordon Cleaver, RAF Pilot

Public domain

During the Battle of Britain, an air war launched by Germany during the summer and autumn of 1940, an RAF pilot, Lieutenant Gordon "Mouse" Cleaver, was returning to his base when a bullet smashed through the plastic covering of the cockpit of his Hurricane fighter plane. Earlier that morning, in the rush to get airborne, Cleaver had forgotten to take along his aviator goggles. His eyes were unprotected as countless minute plastic shards showered him, and as a result he could no longer see. Yet he was able to maintain control of his plane long enough to turn it upside down and safely make a parachute landing. A rescue team spotted Cleaver's landing and immediately took him for emergency medical treatment.

When Cleaver's eyes were examined, the many pieces of plastic shrapnel had completely blinded the left eye and had severely affected the right eye. Major Harold Ridley, an ophthalmological surgeon, examined Cleaver's eyes and ordered that he be transferred to the Moorfields

Hospital, the highly regarded hospital in London specializing in diseases of the eye.

Over the next months, Dr. Ridley performed numerous operations on Cleaver's eye, trying to preserve as much vision as possible. Cleaver returned to civilian life, still having several tiny pieces of plastic embedded in his eye. During his many examinations over the next eight years, Ridley never detected any reaction of the eye to the plastic fragments—such irritation would be expected from any other foreign body. The plastic was simply inert.

The Idea of a New Lens

One of the most common problems that an eye surgeon encounters is cataracts. A cataract occurs when the protein material in the lens starts to cloud up, which diminishes the amount of light reaching the retina. Over time, the cataract grows and becomes opaque, causing the vision to become progressively blurred. Eventually complete blindness ensues. In such a severe case, the lens is completely covered by a white shield that resembles falling white water (i.e., a cataract), hence the name. Cataracts are the most common cause of blindness or near-blindness throughout many parts of the world, especially in older age groups. For several centuries, surgical removal was the only way of treating cataracts, but the outcome for the patient was far from completely satisfactory.

For a week after the operation, the patient had to maintain his head rigidly wedged between two bolsters so that no movement was possible. More importantly, lacking a lens, the patient had blurred vision, since the eye now was unable to focus. To remedy this, the patient had to wear "coke bottle glasses," so named because the heavy thick lenses of the glasses resembled the bottoms of glass Coca-Cola bottles. Even so, vision was still far from normal. Later, contact lenses were developed, and while they replaced the thick glasses, normal vision was still not restored.

A turning point occurred in Ridley's thinking in 1948, when Steve Parry, a medical student watching Ridley remove a cataract, remarked, "It's a pity you can't replace the cataract with a clear lens." Ridley may have recalled a talk with his father, also an ophthalmological surgeon, some years before about the theoretical possibility of removing the cataract and replacing it with an artificial lens. The problem was of course finding the right material that one might use for such a lens. But thanks to his experience with Gordon Cleaver's eye, Ridley had found the suitable material. Later he said that it was Parry's remark that stimulated him to go ahead and try to develop an artificial lens.

Ridley was acutely aware that implanting an artificial lens would bring down the wrath of the conservative ophthalmological community. Eye doctors were accustomed to removing foreign bodies from the eye, but the idea of deliberately inserting one was outrageous. To embark on such a project would require the greatest caution and secrecy to avoid the risks of serious complications on the one hand and the objections of powerful colleagues on the other. They seemed content with the conventional thought that removing the diseased lens and replacing it with imperfect thick glasses was a decent price to be paid for the restoration of some measure of vision.

Ridley knew that using the particular plastic that he had seen in Cleaver's eye, called Perspex, could defeat the rejection process. After several months of working out some details of his plan, Ridley got in touch with John Pike, an optical scientist who worked at Rayners, a manufacturer of optical devices. Pike was an old friend who had earlier worked with Ridley on an electronic device for the examination of the inner eye. Ridley trusted Pike implicitly to keep the project a complete secret. Ridley was so secretive that he held all of his meetings with Pike in his Bentley automobile, which was parked on Cavendish Square in central London.

Pike's expertise was essential for making the artificial lens. After examining the properties of Perspex, Pike thought a higher purity of plastic would be better. Dr. John Holt of the Imperial Chemical Industries (ICI) was able to produce a superior grade of Perspex that Pike named Perspex CQ (Clinical Quality). Perspex CQ proved so satisfactory in Ridley's first patients that it was eventually adopted until even better lenses were developed.

Ridley did not seek a patent on the new lenses, since he wanted to avoid any claim that the artificial lens was a commercial venture. Accordingly, Ridley, Pike, and Holt all agreed not to seek any royalties from any sales of the lenses. They also persuaded Rayners and ICI to set the sale price at slightly less than one pound sterling, despite the losses that they would sustain if the lenses were widely adopted. The entire plan was now complete, except for the crucial step of trying it out on patients.

The First Patients

Ridley thought that the first patient had to be a volunteer who was willing to accept unknown risks of having an artificial lens implanted. He also insisted that his first patient have one perfectly normal eye in case the operated eye developed such major complications that vision was lost entirely. After much screening, he finally identified a forty-five-year-old woman who volunteered to be the first patient.

Ridley performed the first lens implant in St. Thomas' Hospital rather than at the Moorfields Hospital, even though Moorfields was the center for diseases of the eye throughout the United Kingdom. He chose St. Thomas' because he thought that his experiment would arouse less attention from the staff there than at Moorfields. He wanted to minimize the possibility of any leak getting out that he was performing a new, experimental procedure.

On the early afternoon on November 29, 1949, Ridley removed the cataract and implanted an artificial lens with no immediate complications. Following the operation, Ridley asked that the written record of the operation be limited simply to state that the surgery consisted of the removal of the cataract, but not to mention the insertion of the new artificial lens. He decided to wait three months before inserting the plastic lens to be sure that the operated eye was completely quiescent and fully recovered. He reasoned that if there were any adverse reactions after the implant, they would be due to the new lens and not to the previous surgery. As he hoped, the eye showed no reaction to the new lens, and the patient's vision was markedly improved even without any glasses.

Dr. Ridley's penchant for secrecy was well advised, as events would soon show. During 1950 and 1951, he continued to be meticulous in his selection of patients and in his surgical technique. He did allow a few trustworthy supporters to observe the operation, but he swore them to secrecy for two years to be sure of long-term success of the lens implants. Ridley wanted to maintain the secrecy about the first several patients who had excellent results for the next two years, a period during which he could observe the progress of the new lens and detect any complications that might develop. As he explained years later, he was afraid of professional and perhaps even legal consequences if his early results were made public prematurely.

The planned secrecy was broken when one of his patients with an artificial lens implant mistakenly dialed a number in the telephone directory and telephoned the office of Dr. Frederick Ridley, also a Harley Street ophthalmologist, and arranged an appointment. Dr. Frederick Ridley was astounded when he heard the story of the artificial lens. Now that the news was out, Harold Ridley hastened to publish his early results in several leading journals to set the record straight and to gain the priority of discovery.

Opposition Mounts

When the British ophthalmologic establishment learned about Ridley's work, they were furious and accused Ridley of being reckless in his treatment of his patients' visual disturbances. But Ridley persisted in the face of such a highly charged critical atmosphere, which was made even more uncomfortable by the few incidents of complications that some of his patients developed. Yet in most instances, the artificial lenses markedly improved the vision of his patients. Malpractice lawsuits were beginning to be bandied about in England at that time, and Ridley became more and more concerned about being sued. If such a suit were ever brought, he knew that there were plenty of expert eye doctors who would be willing to testify on behalf of aggrieved plaintiffs.

Ridley decided to present his results at the meeting of the Oxford Ophthalmologic Conference in July 1951. He drove his wife and two of his early patients to Oxford, intending to demonstrate his results firsthand. Both patients had nearly perfect vision without glasses. He looked forward to having colleagues examine the eyes of these two prize patients. Ridley was upset when he learned that his presentation was set for the first session immediately after lunch, when some participants were sure to be tardy in returning.

As the session got underway, Sir Stewart Duke-Elder, the greatest ophthalmologist in the United Kingdom, if not in the world, refused to even examine Ridley's patients, and many of his apostles followed, uttering unkind remarks about Ridley. To be sure, a few of the younger, less doctrinaire attendees offered positive comments, and one even congratulated Ridley. But to make matters worse, the chairman of the session brought down the gavel well ahead of the full time allotted for Ridley's presentation. Ridley was deeply hurt and declined the invitation to attend the formal conference dinner.

A few of the younger ophthalmologists were not as strongly bound to orthodoxies as their older colleagues were, and they cautiously promoted the benefits of the artificial lens, but widespread skepticism still prevailed. It was hardly surprising that the great Duke-Elder, Ophthalmologist to the Sovereign and author of the fifteen-volume *Textbook of Ophthalmology*, led the detractors. One of Ridley's younger colleagues summarized the scene: "The initial small enthusiasm for intraocular lenses was followed by scorn and then dismissal. Rather than noting the surprising success of these early attempts, surgeons focused on the probable frequency of failure. A few surgeons even built careers on the effort to consign intraocular lenses to the 'waste bin of ophthalmology.'"

Ridley's first lecture in the United States was well received by the small audience of the Chicago Ophthalmologic Society in March 1952. Seven months later, he returned to Chicago to attend the large national meeting of the American Academy of Ophthalmology. But this time, his talk aroused great hostility and skepticism from leading American ophthalmologists, led by Dr. Derrick Vail, editor of the *American Journal of Ophthalmology*, who wrote, "In spite of Mr. Ridley's remarkably successful run of cases, the operation is of considerable recklessness, its hazards far exceed the little that is gained in the way of ocular comfort to the patient and the questionable advantage of binocular vision obtained at such an obvious risk." Many prominent eye surgeons in the UK and the US followed Vail and Duke-Elder. They treated Ridley's lens with such contempt that it extended to verbal attacks on Ridley's person. The generally poisoned atmosphere, made worse by isolated reports of other surgeons, who had encountered major complications, convinced many to condemn the entire idea. The lens was often referred to as a "time bomb" in the belief that future serious complications were unavoidable.

Here are a few of the comments Ridley received from his audience:

"The operation should never be done."
"The first report was in a lay magazine."
"Would you put one in your son's eye?"
"Dr. Ridley: Why don't you GO HOME?"

The constant stream of nearly universal criticism eventually wore Ridley down. At one point, he fell into a deep depression for which he took medication for some months. But the success of the artificial lens couldn't be stopped, and the idea began to spread throughout the world. Ridley's operation was not only accepted, it was eventually celebrated.

Better Surgical Procedures

Newer surgical procedures and newer materials for the lenses have greatly improved the outcome of patients. One of the great advances in cataract surgery, another instance of an accidental discovery, occurred in the mid-1960s. Up until that time, the removal of the intact cataract required a relatively large incision of the lens capsule, a procedure that sometimes risked uncertain results and complications. The experiences of a young eye surgeon, Charles Kelman, in a dental chair led to the development of a new procedure that avoided that potential danger. While having his teeth cleaned, Dr. Kelman had an inspiration about which he later wrote: "I sat in his [the dentist's] chair, and he reached over and took a long silver instrument out of its cradle and turned it on. A fine mist came off the tip, but the tip didn't seem to be moving. He applied the tip to my teeth, and I felt an exquisite vibration and heard a high-pitched sound. I asked, 'What is that thing?' His reply: 'It's an ultrasonic probe.' This was the moment! This device could liquefy a cataract!"

After several years of painstaking work, Dr. Kelman perfected an instrument called the "phacoemulsifier." (The name is from the Greek *phakia*, meaning "lens," and *a*, meaning "without.") This device allows the removal of the cataract without any large incision in the lens capsule. The tip of the device is introduced through a tiny "keyhole" incision in the lens capsule, and it vibrates many thousands of times per second, disintegrating the cataract, which is then removed by suctioning it out. Using a high-powered microscope, the artificial lens is then slipped in through the keyhole. Since the new lens is folded, it is able to pass through this tiny slit, but once it is inserted, it unfolds and rests in the same position as the natural lens. The incision seals itself off.

Today the operation takes about thirty minutes and boasts a success rate exceeding 99 percent with minimal discomfort to the patient during or after the procedure. Countless millions of cataract sufferers worldwide, including such dignitaries as the late Queen Mum and Nelson Mandela, owe their "new sight" to Harold Ridley. As a memorable footnote, Dr. Ridley implanted an artificial lens in the remaining good eye of the RAF pilot, Gordon Cleaver, late in Cleaver's life.

Indeed, at the age of eighty-four, Ridley had lenses implanted in both of his own eyes. Following the surgery, he remarked, "I am the only man who has invented his own operation." Queen Elizabeth knighted the ninety-three-year-old Ridley in 2000. Sir Harold died a year later, leaving a magnificent legacy for the benefit of mankind. He left an estate of £2.7 million to Pembroke College Cambridge, in addition to £20,000 in memory of an ancient ancestor, Bishop Nicholas Ridley, master of Pembroke, who was burned at the stake in 1555 for his religious beliefs. The Ridley Foundation received £100,000 to continue its work of reducing blindness in the underprivileged. Sir Harold once said, "I would have on my tombstone, 'He cured *aphakia*.'" But he added, "Might someone not ask: 'Who was Mr. Aphakia?'"

Notwithstanding his self-effacing modesty, Sir Harold's work in the field of ophthalmology stands out.

Dr. Robert Drews, a prominent American ophthalmologist, spoke on behalf of all of his colleagues throughout the world when he said, "No matter what brilliant achievements are made in the future, Sir Harold's place in history remains secure." According to Dr. David J. Apple, an eminent eye surgeon and Ridley's fine biographer, Ridley said that he looked forward to meeting Jacques Daviel in heaven. (Jacques Daviel was a French eye surgeon who made important advances in cataract surgery in the mid-eighteenth century, some of which were adopted by Ridley.) Apple added that millions of people who have had their cataractous lenses replaced with the ultraclear vision of lens implants would flank Daviel in heaven.

Comment

The impact that Ridley has had on society throughout the world is immense. Thanks to his intellect, skill, courage, and perseverance, he created an entirely new approach to the prevention of the chief cause of blindness in the elderly. The economic benefits reaped by recipients of the artificial lens are incalculable. But this miracle now needs to be extended to the 20 million people in the developing world who are blind because of cataracts.

Dr. Harold Ridley's discovery of the inert nature of plastic resulted from his recalling the happy accident of being the doctor who saw the plastic shards in the eye of Gordon Cleaver. His wisdom lay in using that observation to perfect an artificial lens using the same inert plastic material that made up the shattered the canopy of Cleaver's plane. Chance has sparked many medical advances. Dr. Ridley's discovery, like those of a number of other discoverers, was heavily criticized initially by orthodox medicine of the time. But great credit is due him for persevering in such an uncomfortable environment for so long a time until the benefits of his discovery were finally realized and brought into the benefit of millions of people.

REFERENCES

Apple, D. J. "Sir Nicholas Harold Lloyd Ridley." *Biog. Memoirs Fellows Royal Society* 53 (2007), 285–307.

Apple, D. J., and D. Sims. "Harold Ridley and the Invention of the Intraocular Lens." *Survey Ophthalmic* 40 (1996), 278–292.

Ridley, N. H. L. "Cullstrand Lecture." *Eur Implant Surg.* 5 (1993), 4–7.

6

Good as New: Dental Implants

A tooth is much more to be prized than a diamond.
                    —*Miguel de Cervantes*, Don Quixote

Introduction

Many people believe that losing a tooth has no important long-term consequences, unless of course the lost tooth is one of the front teeth, the loss of which has obvious cosmetic effects. But the teeth as a whole are designed to work as a team, and when one or more teeth are lost or missing, the teeth adjacent to the empty space created by the loss of their neighbor drift into the open space. This in turn creates an abnormal circumstance that affects the proper function of the teeth to correctly chew food.

Over a long period of time, this change produces an abnormal wear and tear on the unaffected side, which in turn leads to such secondary problems as root decay, loss of fillings, gum disease, and problems with the temporal-mandibular joint. Eventually, the external appearance of the face will reflect a difference due to these effects. The history of dentistry is replete with attempts to replace missing teeth with some sort of appliance that would correct the appearance of the teeth that occurs when one or more other teeth are missing due to disease or trauma. Some of these attempts were successful, but all lacked the sustained ability for secure, long-term fixation to the bone to which they were attached.

Brånemark's Discovery

In 1952, twenty-three-year-old Per-Ingvar Brånemark made an important discovery that would have a profound influence on the way in which missing teeth could be effectively and permanently replaced. Brånemark, who received his medical training at the University of Lund in Sweden, became curious about the detailed functions of the cells in the bone marrow in healing an injury. At the time, there was already a significant accumulation of microscopic pictures of the activities of the various cells of the marrow, but all of these pictures had been derived from the study of dead bone marrow that had been stained with various dyes. Of course, the pictures, colorful as they are, still represented only frozen snapshots of events at a given point in time. Brånemark's interest went beyond these static pictures. He wanted to observe the dynamic changes that were occurring in active, living bone marrow under various conditions.

Brånemark began to search about the methods that he would need to go about viewing the living marrow. He knew that some scientists at Cambridge University were conducting research on the flow of blood using an optical chamber that was embedded in the soft tissue of the ears of rabbits. Brånemark visited the Cambridge researchers and paid close attention to the details of the design of their optic chamber, which was made from tantalum. He left convinced that a design similar to the "rabbit ear chamber" could be developed and used for observing the events occurring in the living bone marrow.

His first problem arose in trying to obtain a source of tantalum, which in the postwar industrial environment proved to be fruitless. Fortunately, in a chance meeting, Dr. Hans Emmaus, an orthopedic surgeon who had been studying various metals that might be used for hip replacements, suggested that he should try titanium, which might serve as well as tantalum.

Heeding this suggestion, Brånemark located a source of titanium and produced a chamber using titanium that was similar to that used by the Cambridge researchers. The chamber, equipped with a microscopic lens, permitted one to observe the relationships of the cells of the marrow with their blood supply. Once he was satisfied, he implanted the chamber in a long bone of a leg of a rabbit. It had taken three years of work to obtain the expertise necessary for making the chamber and for conducting the experiments. Using this chamber, he recorded the results over a period of months, which he published in 1959 as an eighty-two-page supplement in the prestigious *Scandinavian Journal of Clinical and Laboratory Investigation*.

Following the completion of his experiments, Brånemark sought to remove the chamber from the bone, but even using the most strenuous efforts, he was unable to loosen the chamber from the bone in which it had been placed. It simply could not be removed. It had somehow fused with the bone in which it had been embedded. He did not immediately grasp the potential applicability of this finding. Rather he was simply annoyed at losing a valuable piece of equipment. His memory of this curious event would be awakened later by an unexpected turn of events, but at the time, he didn't pursue any further work in exploring the nature of this unusual connection between the titanium chamber and the bone.

In 1960, Brånemark accepted an offer to become an associate professor of anatomy at the University of Gothenburg. Evidently, the university authorities must have been greatly impressed with Brånemark's work in Lund, as summarized in his lengthy supplement of a leading scientific journal, since they offered such a senior academic position to a man who was only thirty-one years of age. This new position gave Brånemark financial support and catalyzed his work, which attracted students who were pursuing the research necessary for receiving advanced degrees.

Stimulated by his earlier work, Brånemark had developed a keen interest in studying the flow of blood though capillaries—i.e., microcirculation. He devised an ingenious method for observing this microcirculation in human volunteers, and using a modified titanium chamber inserted into soft tissues of the arm, he made many new observations on the flow of blood under various conditions. In 1967 and 1968, Richard Skalak, a professor of mechanical engineering on leave from Columbia University, joined Brånemark's research team, and provided new insights in using the principles of mechanical engineering to explain the transformations of blood cells as they flowed through capillaries in human tissue. This work, which resulted in many important publications, represented pioneering studies, as well as launched Brånemark's long-time interest in the field of what is called rheology.[1] Brånemark had fashioned a new titanium chamber to view these activities in humans, and he noted that titanium caused no injury to the skin and soft tissues. During this period, Brånemark returned to the problem posed by his inability to remove a titanium chamber from a rabbit's bone. Professor Skalak also brought an engineer's viewpoints on the design and testing of titanium components and devices that were needed for the research on titanium relationships with bone. In a series of landmark studies, titanium was shown to be structurally and completely integrated into living bone, which resulted in a connection that occurred with a high degree of predictability without any soft tissue

---

1. The term *rheology* was taken from an aphorism of Heraclitus, meaning "everything flows." Rheology is broadly defined as the science of the flow of liquids under specified conditions. Rheological principles are used in many fields other than medicine.

reaction or rejection. The success of experiments in which dogs received dental implants provided support for the idea that such implants could be performed in patients, and that once they were skillfully embedded, they would be long-lasting and free of complications.

This nature of the fusion of titanium and bone was a biological process not previously described, which Brånemark named osseointegration (from the Greek *osteon*, "bone," and Latin *integrare*, "to make whole"). Osseointegration has come to be formally defined as the formation of a direct connection between the implanted material and bone without any intermediate tissue, such as scar tissue, cartilage, or ligament. In other words, osseointegration represents a direct structural and functional connection to the bone to which it is fused. It follows that once the fusion of the implanted titanium with bone is complete and the implant is secure, various objects such as teeth or different prosthetic devices such as artificial limbs can be attached to it.

Dental Implants

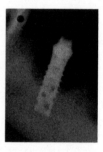

Brånemark was in contact with dentists, orthopedists, and plastic surgeons. In 1963, one of the dentists asked him to see a thirty-four-year-old man named Gösta Larsson. Larsson was born with a cleft palate as well as a deformed chin and mandible (i.e., lower jaw), and he had no teeth on his lower jaw. His life had been miserable due to pain and his inability to chew food. No dental or plastic surgical techniques could repair his speech or correct his other multiple defects. He seemed to be beyond hope. But his dentist did mention that Per-Ingvar Brånemark in Gothenburg was developing some new approaches that might possibly be applicable to Larsson's problems.

After his assessment of the problems presented by this unusual patient, Brånemark proposed that Larsson receive implants in the form of titanium screws, which would serve as connection points to anchor an appliance containing teeth in his lower jaw. After the appliance was put into place and secured with implants, Larsson could normally speak and chew food for the first time in his life. He joyfully adapted to his completely new lifestyle. Larsson died in 2005; after forty years, the implants were still in place.

But following the success of Larsson, Brånemark waited a year, contemplating the pluses and minuses of entering any clinical area with the tools he had developed. There were new challenges in selecting and training dentists or orthopedists in the use of this new technique. Although he was deeply interested in providing new ways of making and attaching new prostheses for amputees, he decided to expand the efforts of his team in the area of dental implants.

Initially, the selection of patients like Gösta Larsson was limited to those with problems so severe as not to benefit from any existing technical procedure. Brånemark now decided to offer dental implants to those whose needs were real but less severe. This decision led to the establishment of the Brånemark Dental Clinic.

Opposition Arises

One might think that the success of these early cases using titanium-containing dental implants would be recognized as a major advance by dentists everywhere. Regrettably, it was not. Practicing dentists in Gothenburg were adamant in their opposition to Brånemark's work. In fact, it would take a period of thirty years before the basic concept of osseointegration would be widely accepted. During the period of the 1950s to 1970s, the scientific literature contained many instances in which the outcomes of placing metals of various kinds in bone led to unfavorable results. These experiences led to the nearly universal belief that that any metal object placed in bone would not only not integrate, but instead would develop a fibrous tissue between the metal appliance and the bone. This fibrous tissue lacked stability and was a target for bacterial infection.

At the time, the general scientific community was simply not prepared to accept the fact that titanium does not form any fibrous layer between metal and bone, but that it actually integrates with the bone in such a way that the fusion of titanium and bone is paramount. Given such an environment, it was difficult to convince the medical world that titanium was a metal that possessed some unique properties, which, although still not fully understood, nonetheless produced a firm integration with bone.

It wasn't long before members of the Swedish dental community, including some academics, rose to criticize Brånemark as well as his team for the work. First of all, Brånemark was not a dentist, nor did he ever assume to be one. To accomplish the results he had presented required the skills from experts in such fields as dentistry, plastic surgery, and engineering. But not even using a dentist was enough for some critics, who regarded him as an outsider of the dental community, and therefore untrustworthy. Matters were made worse when an article describing Brånemark's work appeared in *Reader's Digest*, a popular American magazine for the lay public.

During a lecture, an influential and deeply respected Swedish professor rose and remarked, "I do not trust people who publish themselves in

Reader's Digest." In fact, the article in question had nothing to do with osseointegration, but rather it dealt with Brånemark's earlier work on the microcirculation of the bone marrow. Academics then and now believe that the publication of research results in non-peer-reviewed scientific journals—or worse, in magazines that are directed at the general public—impugn the integrity of the researcher.

But Brånemark could not resist a response to a Professor Björn, who had publically championed a leading brand of toothpicks in which his name was prominently displayed on every packet. He remarked in rebuttal, "I don't trust people who publish themselves on the back of toothpicks." This type of verbal abuse in response to Brånemark's work continued for at least ten years, but it did have the small advantage in highlighting the word and the concept of osseointegration.

Discontent by organized Swedish dentistry came to a head in 1973 with a meeting and a report of the National Ontological Assembly, a leading ontological society. The "Gothenburg method" was condemned, because it involved perforating the membranes lining the mouth, which would make the mouth more susceptible to infection. The proceedings, without the chairman being present, featured many personal attacks, some very vicious, on Brånemark. Alf Ohrman, a highly respected member of the assembly and chairman of the department of oral diagnostics in Gothenburg, rose in anger at the attacks on Brånemark. He knew firsthand of the excellence of Brånemark's work and was appalled at the level of verbal abuse being showered upon him, some of which was personal and vicious. Ohrman threatened to resign in disgust unless the assembly apologized to Brånemark. After several weeks, Brånemark did receive a written apology.

Brånemark and his team faced hostile receptions when they presented some of their results to scientific audiences. For example, at the 1969 meeting of the Southern Swedish Dental Association, Brånemark and his colleagues were the target of shouts like, "You are a humbug, sir—not even a dentist!" These harsh comments and attitudes continued well into 1975, when Brånemark's research team was rejected by almost the entire Swedish scientific community. Brånemark's funds for his studies from the university were discontinued, and there were shouts and whispers for the university that called for the expulsion of Brånemark and colleagues. The attacks were not only against the scientific validity of this new discovery, but they were also leveled personally against Brånemark as well as the members of his team.

First of all, Brånemark, while admittedly not a dentist, had always enlisted colleagues who had essential skills for the projects in dentistry, plastic surgery, engineering, and others when needed. Critics were unfamiliar with the multifaceted nature of such a team, which Brånemark believed to be essential to pursue the goals that he set out.

During his first decade at Gothenburg, Brånemark developed a research program that involved many disciplines that were needed to meet his high standards so that the results were satisfactory. But as the tide of opposition grew, sources of funding began to dry up, and this was magnified by the increasingly tense relations that had developed between the sources of funds and the Brånemark team. This tension was exemplified by the hostility that developed between the Swedish Medical Research Council and the Brånemark team.

Brånemark, with the council's support, had conducted important experiments in osseointegration in beagles, at the end of which the council ordered that the dogs be killed. Brånemark objected vigorously, since valuable long-term information would be lost by not allowing the dogs to live out their natural life spans. Brånemark was reaching a breaking point due to the widespread opposition to titanium implants. A lesser man might have capitulated, but Brånemark didn't waver. Instead, he established a private dental clinic that was not funded by the usual agencies and was outside the national Swedish health-care system. Rather it was supported by such diverse sources as various Swedish industries, private donations, charity, and patients who could afford it.

One of the early patients was a forty-five-year-old man, Sven Johansson, who had no teeth at all and who could not tolerate full-mouth conventional removable dentures. In 1967, Mr. Johansson received eleven implants that secured a nearly full set of teeth. After many years his implants were completely intact. The excellence of Brånemark's private clinic spread, and it began to draw patients not only from Sweden but from other countries as well.

Opposition Fades

In 1974, a group of angry dentists turned to the National Board of Health and Welfare to ask that all procedures involving osseointegration be halted. The board, however, decided to investigate the situation. In February 1975, three highly respected professors of dentistry from Umeå University examined twenty randomly selected patients with dental implants from a total of 165 patients. The professors examined each of the patients and reviewed their complete records and X-rays. Their report, issued in May 1975, fully supported all of the claims that had been made about the implants, and concluded that Brånemark's dental implants were bona fide treatments.

In the light of this favorable report, in April 1976 the National Insurance System accepted dental implants as dental treatments that would be covered by the national health-care system. This was the beginning of the long-term growth of acceptance of osseointegration, which it enjoys today both in dentistry and in medicine.

The Brånemark team could celebrate Brånemark's victory in Sweden and in other European nations, but North and South America remained unconvinced, perhaps because they had not been properly informed. The tipping point came from a forward-looking professor of prosthodontics in Toronto named George Zarb. Zarb had closely followed Brånemark's work, and he became convinced that osseointegration was the key for a bright future for dental implants. Not content to simply read articles about the developing field, Zarb spent the year of 1977 in Brånemark's center in Gothenburg studying osseointegration and confirming and repeating many of Brånemark's original observations. Upon his return to Canada, he published his results in a book entitled the *Republication Study of Osseointegration.*

The year 1982 marked the thirtieth anniversary of the discovery of osseointegration, and an international conference seemed an appropriate time to celebrate Brånemark's discovery. Dr. Zarb organized and played a pivotal role in organizing the conference, which was named the Toronto Osseointegration Conference in Clinical Dentistry. Zarb regarded the conference as an opportunity not only to present the European experiences with dental implants, but also as the means of reaching the dental communities in North America. It was a great success. After the conference, Dr. Zarb was asked his reaction to the idea that osseointegration had finally been accepted. He expressed his feelings in a single word: "Triumphant!" He further explained that it wasn't just a victory for the supporters, but a victory for dentistry, which would enrich the entire community: "It was a giant leap of science . . . we've come a long way." In 2012 the sixtieth anniversary celebration of the discovery was held in five different places: Gothenburg, Avignon, Odessa, Hamburg, and Toronto.

Dental Implants in the United States

One of the most important outcomes of the Toronto conference was the interest shown by representatives from the Mayo Clinic. This prestigious clinic became one of the first locations in the US to train oral surgeons and periodontists in dental implantation. The Academy of Osseointegration was formed in the United States, and the American Dental Association approved the type of dental implants originally used by Brånemark. Initially, dental implantation was reserved only for patients who had lost all their teeth. The treatment was viewed as a last resort for those who had no other options.

But as the high success rate of osseointegration and the relatively low rate of complications accumulated, prosthodontists began considering implants for less severe cases. This began with implant-supported bridges and eventually single tooth implants. There were setbacks in the 1980s as various implant designs and placement techniques were tried. Some worked and some didn't. The treatment was far from standardized in its early days. Even now, there are many Brånemark-type implants on the market, and no single method of implantation is considered the best. Dentists now have a lot more data and experience to draw on in selecting the ideal type of implant and the best placement method to ensure success for each patient.

Nearly a half million dental implants are placed each year. The procedure is considered a routine part of dental restoration. The aesthetic and functional results are considered superior to any other options for replacing a lost tooth or teeth. The success rate for osseointegration is 90 to 95 percent, a remarkably high rate for any oral surgery procedure.

Brånemark's Honors and Awards

Over the span of six decades of research, Professor Brånemark has received many awards and honors, including the Swedish Society of Medicine's Söderberg Prize (the mini-Nobel Prize) in 1992 and the Swedish Engineering Academy's medal for technical innovation. In addition, he has received the Harvard School of Dental Medicine Medal; thirty honorary academic positions, including an honorary fellowship of the Royal Society of Medicine (London) and an honorary doctorate (2003) from the European University of Madrid; and the 2011 European Inventor Award for lifetime achievement.

REFERENCES

Brånemark, P. I. "Vital microscopy of bone marrow in rabbit." *Scan J Clin Lab Invest.* Suppl. 38 (1959), 1–82.

Brånemark, P. I. et al. "Osseo integrated implants in the treatment of the edentulous jaw. Experience from a 10-year period." *J Plastic Reconstruct Surg.* Suppl. 16 (1977), 1–132.

Skalak, R., and P. I. Brånemark. "Deformation of Red Blood Cells in Capillaries." *Science* 164 (1969), 717–719.

Part III
Drugs that Affect the Blood

7

Vitamin K: Began with a Bleeding Chick

Koagulation-Vitamin.
> —*Henrik Dam*

Introduction

Henrik Dam was born in Copenhagen in 1895, the son of a pharmaceutical chemist. After completing the six-year program in chemistry at the Polytechnic Institute in Copenhagen, where he earned the MS degree, he joined the Royal School of Agriculture and Veterinary Medicine, rising to the rank of professor. Dam's research interests focused on the biological importance of sterols, a group of compounds chemically related to cholesterol, as well as to vitamin E and to gallstone formation. Dam became deeply interested in the synthesis of cholesterol in various species. It was already known that various animals, such as rats, mice, and dogs, could synthesize cholesterol, but there was uncertainty about the ability of chicks, based on previously published reports that seemed to show that chicks could not thrive on a diet from which the fats and sterols had been removed by extraction. Such results were clouded by the possibility of having deprived the chicks of the fat-soluble vitamins A and D.

Dam set out clarify this confusion by feeding chicks a chemically pure diet that was free of all sources of fat including sterols but was fortified by the addition of vitamins A and D. He found that a considerable part of the cholesterol in the newly hatched chick is the result of transfer from the egg yolk, but that it disappears during the first two or three weeks as cholesterol is formed in increasing amounts as body weight increases. These studies clearly proved that the chicks could make cholesterol. But an unexpected symptom began to show up in chicks that were kept on the diet for more than two or three weeks. They developed hemorrhages under the skin and in their muscles. In pursuing this new finding, Dam excluded any role of vitamin C or the lack of fat in the diet. Using the hemorrhagic chicks as his assay, Dam determined that the unknown factor was widespread in many animal organs and plant materials. Large amounts of cereals and seeds also prevented the hemorrhages.

Dam now suspected that he was on the track of a new and unknown factor that was essential for the coagulation of the blood. After several more years of work, he was satisfied that his factor was a new discovery, and he named it vitamin K, after the Danish and German word *Koagulation*. Professor H. F. Almquist, working at the University of California, was also studying the hemorrhagic disease of chickens. He discovered that chickens fed on sardine-meal as the source of protein developed the condition, while those fed on meat-meal did not. Adding alfalfa to the sardine-meal protected the animals against the hemorrhages. Almquist found that ether extracts of the meat-meal prevented the hemorrhages of the chickens fed the sardine-meal. He also showed that the fat-soluble preventive factor was not identical to any of the known fat-soluble vitamins.

The factor was produced in feedstuffs by subjecting them to bacterial action. Almquist correctly concluded that intestinal bacteria were making vitamin K from the fat-containing diets of grains and seeds. He showed that the difference between the two protein sources was that the meat-meal had a slower passage through the intestine, and this gave the bacteria in the intestinal tract time enough to produce the protective factor. Almquist summarized all these results in a paper that ended with the announcement that he had discovered a new vitamin.

In a curious twist, Almquist's paper was submitted to the American journal *Science*, which rejected it because of a paper that had been submitted by Romayne Cribbett and John Correll, who were working in a different department of the same university on the same Berkeley campus as Almquist. Cribbett and Correll's paper presented a quite different view of the cause of the hemorrhagic disease in the animals, a view that directly conflicted with Almquist's views. The university authorities stepped in and enjoined both parties from submitting any further reports for publication until this controversial matter could be resolved. The resolution took several years, but with Dam's full support, Almquist's paper was finally published in the British journal *Nature*. This settled the dispute on the campus, but left Almquist in second place for claiming priority of the discovery of a new vitamin and a potential candidate for the Nobel Prize, which was awarded to Dam in 1943.

Later, Dam, who had previously visited Almquist's laboratory, vouched for the accuracy of Almquist's work and sent a gracious letter to him in which he commented that Almquist had so "narrowly missed" being the first to report the existence of vitamin K.

In both Dam's and Almquist's laboratories, work proceeded on the methods of assay and the properties and purification of the vitamin. Dam and his colleague Fritz Schonheyder discovered the mechanism by which the vitamin exerts its effect. During the process of blood coagulation, the formation of fibrin requires the action of thrombin that is formed from prothrombin. They found that vitamin K increases the production of prothrombin and therefore can counteract bleeding from any disorder in which prothrombin deficiency occurs. This was an important advance since clinical disorders in which prothrombin levels were low could now be treated with vitamin K. Later, Harold Campbell and Karl Paul Link found that vitamin K inhibited the action of dicumarol and warfarin, two agents that they had developed that delayed the clotting of blood.

Further Studies

During the 1930s, Dam studied the medical applications of vitamin K. Through clinical colleagues, he learned that vitamin K was instrumental in correcting a number of bleeding disorders that were due to low levels of prothrombin. The vitamin proved essential in preventing hemorrhagic disease of newborn infants. In healthy individuals the bacteria in the intestinal tract produce vitamin K, which is absorbed into the blood. But in the newborn and especially in the premature infant, an interval of several days is required before the intestinal flora is fully developed.

Serious hemorrhages may occur in this interval, but they can be prevented by the prompt administration of vitamin K after birth. Dam now sought to define the chemistry of vitamin K, and in this he gained the assistance of Professor Paul Karrer, a famous chemist who had won the Nobel Prize in Chemistry in 1937. By 1939, Karrer and Dam had isolated the vitamin in pure form and determined its molecular structure. Simultaneously, Edward Doisy in St. Louis isolated the vitamin from alfalfa and determined its structure. He named this substance vitamin K_1. Doisy went on to discover and determine the structure of the compound that was produced by the intestinal bacteria and named it vitamin K_2.

Shortly after World War II began in Europe, Dam came to the United States on a lecture tour. When Germany invaded Denmark in April 1940, Dam decided to stay in the United States throughout the war. He had academic appointments during that period from the University of Rochester and subsequently from the Rockefeller Institute for Medical Research (now Rockefeller University). He was offered a professorship at the College of Advanced Technology in Copenhagen in 1941, but could not travel back to Denmark at that point.

Dam was in New York in 1943 when he learned that he had been awarded the Nobel Prize in Physiology or Medicine along with Edward Doisy. Dam's citation was for "his discovery of vitamin K" while Doisy was recognized for "determining the chemical structure of vitamin K." The Nazis censured the Danish media and withheld any mention of the Dane who had won the Nobel Prize. But Dam's award was not overlooked in the United States. A ceremony was held at the Waldorf Astoria hotel in New York for Dam and four other Nobel recipients in 1943 and 1944. Dam received congratulatory telegrams from many notables, including the Swedish king and President Franklin D. Roosevelt. The actual Nobel Prize medal and the diploma did not arrive in the United States until May 1945. In 1946, Dam returned to Denmark as professor of biochemistry and director of the Danish Lipid Research Institute. He died in Copenhagen in 1976 at the age of eighty-one.

Comment

Dam's discovery is a near perfect example of the type of discovery that Horace Walpole had in mind in his entertaining tale about the travels of the Three Princes of Serendip (Sri Lanka). Walpole relates that the princes were always making discoveries by accident and sagacity, from which Walpole coined the word *serendipity*. In strict modern usage, serendipity has two components, the accidental discovery and the sagacity (i.e., wisdom) to recognize its value. Henrik Dam's discovery fits that mold perfectly. His basic research interest was in cholesterol metabolism, and in his first experiments, he sought to determine if chickens needed an external supply of fat to make cholesterol. He found that they could synthesize cholesterol on a fat-free diet, but that the diet he used resulted in hemorrhagic bleeding. Having excluded all of the then-known causes of such bleeding, he dropped his studies of cholesterol and concentrated on a new path of study that led to the discovery of vitamin K.

REFERENCES

Almquist, H. "The early history of vitamin K." *Amer J Clin Nutr.* 28 (1975), 656–659.

Dam, H. "The discovery of vitamin K, its biological functions and therapeutically application." Nobel Lecture, Nobel Prize in Physiology or Medicine, 1946.

Dam, H. "Medical Aspects of Vitamin K." *Lancet* 63 (1943), 353.

Dam, H. "Vitamin K, its Chemistry and Physiology." *Advan Enzym.* 2 (1942), 285–296.

CHAPTER

8

Dicumarol and Warfarin:
Began with a Bleeding Cow

Vat vill he haf on Monday? More dead cows!
—Eugen Schoeffel, German laboratory assistant

Ed Carlson's Bleeding Cows

In February 1933, Ed Carlson, a Wisconsin farmer from Deer Park, some 190 miles from Madison, saw his cattle herd falling one by one as victims of a disease in which they developed large hemorrhages in their muscles and under the skin. In December, he lost two young heifers, and in January one of his old cows developed massive bleeding that was fatal. Two young cows had died on the previous day, and the bull was bleeding from the nose. Their blood simply would not clot, and they were dying from blood loss. Carlson's veterinarian was uncertain about the cause, but guessed that the bleeding was due to some unknown vitamin deficiency as a result of the cattle eating spoiled sweet clover.

Carlson was not satisfied with the veterinarian's opinion. Unhappy with the uncertainty, he wanted more information and thought the best place to go would be the Wisconsin State Agricultural Experiment Station, which was part of the University of Wisconsin in Madison. He loaded his truck with a frozen dead heifer, a milk can filled with unclotted blood, and a hundred pounds of sweet clover, and he set off in a howling blizzard to drive on snow-covered roads to Madison, 190 miles away.

Karl Paul Link

Arriving in Madison, Carlson found that all of the doors of the Agriculture Building were locked, since it was a Saturday. Not to be undone, he walked to the next building, which luckily was the home of the biochemistry department. Wandering through the deserted halls, Carlson ran into Karl Paul Link, a young agricultural chemist, who was accompanied by one of his graduate students, an impetuous German immigrant named Eugen Wilhelm Schoeffel, whose English was poor but whose German education had been excellent. At the time, Link was trying to develop a strain of sweet clover that was free of coumarin. Coumarin was a well-known substance that was responsible for the sweet smell of freshly cut clover, but had a bitter taste that made the clover less attractive to cattle.

Link and Schoeffel stopped to help Carson, who began to tell his tragic story that was threatening him with financial disaster. Link had heard about the "hemorrhagic agent," a name commonly used by some veterinarians for the unknown agent. Link told the distressed Carlson that the only things that could be done were either to avoid feeding the spoiled hay or to transfuse the sick animals. Carlson responded that he had no resources to implement either of these suggestions. Link added that future research would eventually find the cause and the cure for the disease. The disheartened Carlson returned to his northwestern Wisconsin farm, again encountering the heavy blizzards along the way.

After Carlson left, Schoeffel, storming back and forth in the laboratory, shouted in his heavy accent, "Vat ad hell, a farmer shtruggles nearly two hundred miles in dis sau-wetter, driven by a shpectre and den has to go home vit advice: 'Get some good hay or transfuse.' Ach! Gott, how can you do dat ven you haf no money? Vat vile he fined ven he gets home? Sicker cows. Vat vill he haf on Monday? More dead cows! He has no udder hay to feed and he can't buy any. And if he loses de bull, he loses his seed. Mein Gott! Mein Gott!"

As Link was leaving for home, Schoeffel grabbed him by the shoulders and said directly in his face: "Before you go, let me tell you something. Der is a deshtiny dat shapes our ends, it shapes our ends, I tell you!" It was indeed a prophetic statement. Link was impressed with Schoeffel's fairly close quotation from *Hamlet*.

Sweet Clover Disease

In the first decade of the twentieth century, European sweet clover had been planted on vast areas of the Northwest plains, since the overfarmed soil and the harsh weather conditions in that region no longer supported the traditional animal feed crops. Sweet clover was grown widely as a hay crop in the northern United States and in Canada. Its use as hay was widespread in the 1920s, but following a series of wet summers, an epidemic broke out in cattle of the "bleeding disease." Some veterinarians thought the new disease was some form of bleeding related to some unknown vitamin deficiency, such as the clotting disorder that was seen in vitamin K deficiency. Still others were convinced that it was due to a change in the sweet clover cattle feed, but they were mystified as to any further details.

Veterinarians were able to trace the cause to sweet clover hay that had been improperly cured and had become infected with molds. When sweet clover is put up as hay, the lush foliage is quite susceptible during cutting and curing for contamination by molds, especially under wet conditions. Hundreds of farmers faced economic disaster when their herds were widely affected with fatal hemorrhages. Switching the feed of affected cattle from sweet clover to another feed was effective, but since sweet clover was the mainstay of cattle feed, all non-clover silage was in short supply and costly. Transfusions of fresh blood from unaffected cows were also effective, but only if they were given early in the course of the disease—and the transfusions were costly and hard to manage.

An authoritative report from the State Agriculture Station in North Dakota reported, "It is not likely that the coumarin is the cause of the injury from damaged sweet clover unless it results from decomposition. The fresh green plant and well-cured hay do not cause characteristic injury of the damaged product." Veterinarians discovered that the level of prothrombin in the blood, an essential clotting factor, fell progressively as the disease progressed. When it reached a critically low level, hemorrhages occurred. All of the other factors needed for blood clotting were normal.

The Search Begins

Link's chance conversation with Carlson set him off on a personal crusade aimed at finding the cause of the fatal bleeding disease. He began by extracting spoiled sweet clover using the prothrombin test in rabbits as the measure of the effectiveness of each step in the purification process. He first focused on coumarin as a starting point, since he was already familiar with its chemistry, but he soon found that it did not cause a reduction in prothrombin levels.

But one of Link's students discovered that formaldehyde, which is produced in the spoiled clover, produced a derivative of coumarin, which they called coumadin and later renamed dicumarol. Unspoiled clover did not produce dicumarol, but when coumarin was added to alfalfa hay and allowed to spoil, it produced a dramatic fall in the prothrombin level.

The general procedure for identifying an unknown chemical is to isolate it, purify it, and finally crystallize it in pure form. The first two steps were uncomplicated, but crystallization proved difficult. Harold Campbell, a graduate student of Link's, worked day and night for many weeks trying to crystallize what would become dicumarol. Very early on a June morning in 1939, Campbell, peering through a microscope, saw the beautiful, oblong crystals of dicumarol. When Link arrived in the laboratory later in the morning, Campbell was sleeping on an old couch nearby. His assistant was awake but drunk. He excitedly told Link, "I'm celebrating, Doc, Campy has hit the jackpot."

Campbell successfully repeated the crystallization and then waited a few days before giving Link a vial containing all of the 6 milligrams of pure crystalline "hemorrhagic factor," adding the understated comment, "Here is your H.F." Preliminary analyses of dicumarol yielded some important facts about its structure. Link had long suspected that vitamin K, the lack of which caused hemorrhages in chickens, and the factor that caused sweet clover disease were closely related chemically. This idea was indirectly validated when vitamin K proved to be effective in overcoming the effects of dicumarol in the test animals. Campbell went on to demonstrate that the hemorrhagic agent was unique and was not identical to any of the known, naturally occurring substances in sweet clover.

Further studies required a larger amount of dicumarol in pure form. Arnold Stahmann and Charles Huebner, two of Link's students, undertook a large-scale extraction of spoiled sweet clover using large oak barrels. After four months of work, they isolated 1.8 grams of pure dicumarol, enough to continue their further studies. They set about to map the chemical structure of dicumarol using some of the 1.8 grams of material that they had previously extracted.

Campbell had determined that the dicumarol molecule contained nineteen carbon atoms, but Stahmann and Huebner could account for only fourteen, as they tried to reconstruct the complete molecule. In April 1939, they set out to find the missing five carbon atoms. But more than that, over the next three days, they were able to synthesize the entire molecule. Their synthetic dicumarol matched exactly all of the properties of the naturally occurring substance with which Campbell had worked. Furthermore, the ability to make dicumarol from inexpensive starting materials opened the door for mass production and commercialization. Link and his students went on to develop a hundred chemical relatives of dicumarol, called analogues, but only one, number 42, proved to be more potent than dicumarol. (More about number 42 later.)

After all of the animal tests had been completed, Link and colleagues were ready to collaborate with clinicians in trials of dicumarol as an anticoagulant. The initial tests occurred at the Wisconsin General Hospital and later at the Mayo Clinic. They indicated the potential of dicumarol for control of clotting in patients. The fact that vitamin K was able to counter the action of dicumarol reassured the doctors that they had an agent to counteract any untoward effects of dicumarol. Link assigned the valuable patent rights to the Wisconsin Alumni Research Foundation (WARF).

Clinicians were eager to use dicumarol in patients with blood clotting problems in the leg veins. In 1942, proposals were made to use dicumarol to reduce the clotting of blood in the coronary arteries of patients with coronary thrombosis. Even though this suggestion was viewed as highly controversial, some early trials were successful, and the use of dicumarol for heart attacks became widespread, only to fall out of favor in the 1950s and 1960s in favor of another anticoagulant that Link and his team would develop.

The Birth of Warfarin

In 1945, Link contracted tuberculosis and spent eight months in a sanatorium. During his stay, he occupied his days meticulously reviewing all the laboratory data available on each of the hundred coumadin derivatives that had been synthesized. He also read everything that he could find on the subject of rat poison. Since dicumarol was only weakly effective against rats, he thought that one or more of the analogues might be superior as a rat poison. Link dismissed number 42 of the analogues because he thought that it was far too toxic to ever be used for clinical purposes, and he concluded that its patent would have no use.

But a key worker on this project was Arnold Stahmann, a shy, introverted chemist and the complete opposite of the flamboyant Link. In Link's absence, Stahmann, who was in charge of the laboratory, saw a brighter future for number 42, and he proceeded to patent it, listing the co-discovers as Link, Stahmann, and Ikawa Miyoshi. The name of the new compound was warfarin, from the initials W-A-R-F (Wisconsin Alumni Research Foundation).

Following Link's return to the laboratory, an uncomfortable hostility broke out between Link and Stahmann that centered on who should be credited with the discovery of the new anticoagulant. The dispute nearly came to blows between the two. Link ordered Stahmann to cease all further work on anticoagulants. Stahmann recorded his bitterness toward Link in his private diaries after he retired. Warfarin was first promoted in 1948 as a rodenticide. It was potent enough to kill rats that ingested multiple amounts that were given in warfarin-containing bait or water, even when they were also receiving vitamin K. Warfarin differed from all other

rodenticides in that its action was delayed, a finding that was not appreciated until later, when it became the most used poison for rats and for other
small animal pests. Warfarin acts as an inhibitor of the action of vitamin K
by blocking the normal process of blood clotting, thereby leading to fatal
hemorrhages in rats.

Warfarin is more potent than dicumarol, but like dicumarol, its actions
can be countered by vitamin K, but only in very large doses. At first clinicians were afraid to try warfarin on patients because of its high toxicity in
animals. Link was amused that the doctors had gladly accepted "my cow
poison" but were reluctant to use "my rat poison." The relatively low toxicity of warfarin in man was clear when a young navy recruit attempted to
commit suicide by taking a large amount of warfarin. He survived and his
recovery was uneventful.

Clinicians now readily switched from dicumarol to warfarin, and it
became the drug of choice for treating clotting disorders of veins and arteries. It was also useful in treating heart attacks by preventing the clotting of
the blood in the coronary arteries. Warfarin achieved wide publicity when
it was given to President Dwight D. Eisenhower as part of the treatment of
his heart attack in 1955.

Link: The Man

Karl Paul Link was an unusual, charismatic person who attracted many
students to his laboratory and to his home, which was always open to all.
He thought of himself as a "man of the soil." Originally, he had planned
to pursue a medical career, but instead he devoted fifty years to chemical
research and contributed far more to medicine than if he had become a
physician. Link had a fine reputation as a carbohydrate chemist and brilliant teacher. He inspired his students with his special gift of being able to
present complex ideas in simple and easily understandable terms. He gave
his graduate students significant freedom in pursing their research projects, withholding advice unless requested. Many of his graduates went on
to outstanding careers in industry and academia.

On campus, Link had a distinctive appearance, dressed in his worn
black herringbone jacket, flannel shirt, large-knotted tie, and wide-
brimmed hat. He could have easily been mistaken for a successful
farmer. Within university circles, Link had the reputation of being an
outspoken iconoclast and a vocal critic of university policies when they
went counter to his and his students' interests. On one occasion, he
severely chastised the university trustees in the newspapers for under-
funding one of his favorite projects. The trustees responded by formally
censuring him. Link's political leanings were decidedly leftish. He was

the faculty advisor to several left-wing student groups, one of which was Marxist in its beliefs. He donated his own money for legal defenses of students who shared his political views.

He was adamant in his opposition to the atomic bomb nuclear tests and the "witch hunt" being carried out by the House Un-American Activities Committee. He refused to participate in a university-wide photo identification program by refusing to be photographed. Link was generally an amiable person, but he also had a violent temper when riled up. Link hated a fellow faculty member, Harry Steenbok, perhaps because of Steenbok's fame from his work on vitamin D and because he had held a number of lucrative patents. During an angry confrontation, Link grabbed Steenbok by the neck and was ready to land blows before being stopped by onlookers.

Link opened a Harvey Lecture in New York with the following: "My pleasure and surprise must also be indicated at finding so many interested in spoiled sweet clover. I had no idea that Manhattan and its environs harbored so many country gentlemen." Link died in 1978 at the age of seventy-seven. A colleague wrote of his scientific achievements: "He never ceased to wonder, he kept on trying and his research was directed toward doing mankind some good."

Comment

The discovery of potent anticoagulant drugs that found wide use in medicine occurred by a remarkable series of chance events involving Ed Carlson, the distraught famer, and his chance meeting with Karl Paul Link and Eugen Schoeffel. Link's standard suggestion of changing the feed to a non-clover source or transfusing the bleeding cows fell on Carlson's deaf ears. He couldn't afford either. Deeply discouraged, Carlson left to face the long trip back to Deer Park in the unrelenting weather.

Link's interest in Carlson's problem may in part have grown out his sympathy for the poor man's plight, but it also may have been aroused by his current research, which was concerned with the chemistry of coumarin, a component of sweet clover. Link could have walked away from the meeting with Carlson without changing the direction of his work. But he didn't. In the words of Salvador Luria (Nobel laureate, 1969), it was another instance of "the chance observation falling on the receptive eye."

Arnold Stahmann, Link's old collaborator, died in 2000 at the age of eighty-six. Following his death, his daughter released an autobiographical essay in which he had written, "My work in this area helped make tissue transplants and artificial heart valves possible yet, my work is not widely known, for Professor Link later turned against me and has taken most of the credit for my research." Stahmann deserves a great deal of praise for the many discoveries that he made during the decade that he spent with Link. Leaving Link's laboratory, Stahmann was initially appointed assistant professor of biochemistry at Wisconsin, but continued through a long and distinguished career to the rank of professor.

REFERENCES

Burris, R. H. "Karl Paul Link Biograph Memoir." *Nat'l Acad Sci D.C.* 65 (1994), 175–195.

Link, K. P. "The Discovery of Dicumarol and its Sequels." *Circulation* 19 (1959), 97–107.

Part IV
Drugs that Fight Infection

Introduction

Smallpox, blood poisoning, meningitis, pneumonia, and tuberculosis—these are diseases that once evoked terrifying scourges that spread across the land in mysterious ways. In 1870, Louis Pasteur, having dispelled the idea of spontaneous generation of germs, went on to demonstrate that microbes were responsible for fermentation and to suggest that microbes were the cause of many diseases. Robert Koch expanded this concept by linking specific bacteria to specific diseases and in the process paved the way for developing cures. The body reacts to an invasion of a microbe by marshaling two interrelated defenses. White blood cells respond immediately in an attempt to destroy the offending microbe, and later a specific antibody is produced by the immune system, which aids the action of the white blood cells in disposing of the responsible microbial agent.

Edward Jenner in 1796 was the first to understand that a "tamed" infectious agent could be used to prevent a subsequent infection in other people by producing an antibody against the infecting microbe. In other words, the body's immune system, once stimulated by a mild dose of a given germ or a modified germ, produces an antibody as part of its natural defense. Such an antibody confers protection against any later larger invasion by the same virulent microbe. Pasteur applied this concept to animal and human diseases with great success. For a time this immunological approach was seen as the Holy Grail for the prevention of all infectious disease. But it all started with Jenner.

When an overwhelming invasion of pathogenic agents occurs, the body's defenses often can't cope and death ensues. Could a drug be found to directly attack and kill the microbes themselves? A succession of early discoveries that were the offspring of the dye industry opened the minds of physicians to the new idea that drugs taken internally could be an effective means for curing infections. The discovery of penicillin launched a new age. The word *antibiotic* is derived from Greek roots: *anti*, meaning "against," and *bios*, meaning "life." Professor Selman Waksman coined the word in 1942 to describe any substance produced by a microorganism that antagonizes the growth of other microorganisms. Penicillin was the first antibiotic to be discovered. It was a drug from a natural source and not the result of chemical wizardry. The discovery of streptomycin extended the antibacterial weapons through its ability to kill the dreaded tubercle bacillus. The third such major discovery was cephalosporins.

C H A P T E R

9

Smallpox: The Cure of a Lesser Evil

The most terrible of all of the ministers of death.

—*T. B. Macaulay*

Introduction

Smallpox has been the scourge of mankind for centuries, due to its highly contagious nature and its devastating consequence on its victims. The virus owes its easy transfer from person to

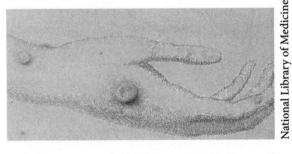

National Library of Medicine

person because it is airborne and because it takes only an exceedingly small amount of virus for a person to breathe in order to become infected.

During the first ten days of the incubation period, the victim feels few if any symptoms. But then he or she experiences a sharp rise in temperature, the onset of a severe backache, and a widespread rash consisting of numerous small red spots. These lesions soon are transformed into blisters, called pustules, which grow in size and rupture, emptying their contents of fluid.

73

The skin splits horizontally, which tears the dermis from its underpinnings and from the upper layer above. The splitting is excruciatingly painful. Pustules may attack the eyes and blindness often results. If the victim survives, the pustules become hardened, bloated, pus-filled sacs the size of peas, which are spread over the entire body. These pustules mature into scabs and fall off, leaving many ugly, disfiguring scars. Death occurs due to cardiac or respiratory failure.

Some people develop an extreme case of smallpox, which is loosely called "black pox," of which there are two different forms. In one form, the skin remains smooth and free of pustules, but it turns dark, presenting a char-like picture from which skin may slips off in sheets.

In the other so-called hemorrhagic form, thick uncoagulated blood runs from the mouth and other body orifices. This form is nearly always fatal. In such cases the virus disintegrates the linings of the throat, the stomach, the intestines, the rectum, and the vagina. Fatal smallpox can destroy the body's entire skin—both the exterior skin and the interior membranes that line the passages of the body.

Some History of Smallpox

The smallpox virus first infected the human race about 10,000 years ago, probably in Egypt. The mummified head of the Egyptian pharaoh Ramses V bears the pockmarks of smallpox. The disease periodically decimated Europe and the rest of the world in the ensuing centuries. The colonists brought the disease to the New World, where it had a devastating effect on them as well as on the Native Americans. Throughout the intervening centuries, smallpox spread across every region of the earth, killing as many as half a billion people.

The Chinese recognized the disease more than 3,000 years ago. They knew that survivors of the disease developed a lifelong resistance to reinfection. During the Ming dynasty, they developed their own method of preventative immunization, which consisted of ground-up smallpox scabs that were gently blown through a tube into the right nasal passage for males and the left for females. A mild form of smallpox usually developed, but it conferred immunity against subsequent exposures to the disease. An English traveler returning from the Orient spread the news of this practice to the London physicians, but his message was ignored. It would not be the first time that orthodox medicine ignored a promising new idea.

Lady Mary Worley Montague is a memorable figure in the history of smallpox. She was nobly born, was gifted with high intelligence, and would become a famous writer of the eighteenth century. When her husband, Sir Edward Worley Montague, was appointed as British ambassador to the

Ottoman Empire in 1716, Lady Mary accompanied him to Istanbul. Early in the next year, she contracted a severe case of smallpox from which she recovered but was left with horrible scars on her face and body.

Later in the same year, Lady Mary's pregnancy had proceeded to term, and an English physician, Dr. Charles Maitland, who was assigned to the embassy, attended her delivery. Dr. Maitland, in need of an assistant, by a stroke of good luck chose Dr. Emmanuel Timoni, a Turkish physician. Dr. Timoni could not help but notice the scars on the once-beautiful face of the proud new mother.

Timoni was well aware of an Arab practice for the prevention of smallpox. The procedure consisted of rubbing small amounts of the liquid obtained from smallpox blisters through small incisions on the arm of a healthy person. Timoni had published a tract in English outlining the details of the practice. The London physicians of the day completely ignored his publication, perhaps because they looked down on all such Arabic medical practices. But they did accept a procedure called variolation, which was an English word that was used to describe the Arabic procedure. The name was derived from the scientific name of the smallpox virus, variola. (The word *vaccination* would come later.) Variolation used live smallpox virus, and it usually resulted in a slight temporary illness. However, if the dose of virus was excessive, full-blown and even fatal smallpox resulted. This was a constant threat, since there was no way of knowing the precise amount or potency of the variolated virus that was given.

Lady Mary asked Timoni to variolate her older son, and upon her return to London, she had her infant daughter variolated. There were no complications, and the mother felt that her two children were protected from smallpox. She was so impressed that she persuaded several highly placed physicians of the Royal Society of Medicine as well as some of the London newspapers of the merits of variolation. She soon became a distinguished and articulate advocate for variolation. She even convinced Princess Caroline to have her two daughters variolated, a step that conferred royal approval for variolation. But despite Lady Mary's huge efforts, variolation faded when an increasing number of variolated people died of smallpox due to having received excessive doses of the virus. In Lady Montague's eighteenth-century England, smallpox was rampant, with a fatality rate of 40 percent. The basic question remained: How could smallpox be prevented?

Edward Jenner

The answer would come came from a man, Edward Jenner, who was born in Berkeley in 1749, the son of the local vicar. After a thorough grounding in the classics, Edward decided at the age of thirteen to become a physician, and he was apprenticed to Mr. Ludlow, a surgeon of Sodbury near Bristol. On one occasion during his six years of apprenticeship, Jenner happened to overhear a comment from a young milkmaid who had come to Mr. Ludlow for medical advice. When smallpox was mentioned as being prevalent in the region, the young woman exclaimed, "Thank God, I cannot take it, for I have had cowpox." This offhand remark made a deep impression on young Jenner, who resolved to explore the young woman's comment at some future time.

Cowpox was a mild disease that affected the udders and teats of British cows. Milkmaids often developed a mild pox-like disease that lasts several days. The effect of cowpox as a preventative for smallpox had long been part of the folklore of the countryside in Germany as well as in England. In 1769, a German farmer, Jobst Bose, wrote, "I am reminded of the attacks of cowpox, which to this day, milkmaids are subject. In this country those who have had the cowpox flatter themselves to be entirely free from all danger of getting smallpox. I have heard this statement made by entirely reliable persons."

Following his medical apprenticeship, the twenty-one-year-old Jenner moved to Berkeley, where he settled and took up his practice as a country practitioner. Shortly after his return, he courted and became engaged to a wealthy Berkeley lady, but the wedding never occurred for reasons that remain completely obscure. Jenner wrote to John Hunter of his deep disappointment, and Hunter wrote back advising Jenner to study the hedgehog's habits to relieve his love sickness. A few years later, Jenner married, and he and his bride, Catherine, settled in Berkeley.

Once in the country, and having been stimulated by Hunter's deep interest in nature, Jenner took up a study of the habits of the cuckoo bird, since his medical practice was still small enough to afford him leisure time. The cuckoo bird, which never builds a nest of its own, fascinated Jenner, preferring to deposit its eggs in the nest of a hedge sparrow. Through diligent observation, Jenner learned that the cuckoo arrives in mid-April, lays its eggs a month later in the sparrow's nest, and then departs in July before the eggs are hatched.

The sparrow acts as a foster mother and incubates the eggs for two weeks. Once hatched, the nestlings require several more weeks in the nest before they can fly. By this time, the parent cuckoos have long since departed without attempting to feed or care for their newly hatched offspring. The sparrow continues her role as foster mother and feeds the nestlings. Jenner made another curious observation.

The strongest cuckoo nestling, although blind, gropes around the nest searching for other nestlings or unhatched eggs. Using the tips of its wings for guidance, this cuckoo nestling ejects unhatched eggs and all of the other nestlings from the nest. When Jenner tied lead weights to the legs of the nestling, it was unable to toss out any of the other nest inhabitants. Jenner reported these observations to the Royal Society of London and was subsequently elected as a fellow.

Jenner's Initial Experiments

Jenner continued to be intrigued by the story of the milkmaids. In 1796, cowpox appeared on a farm near Berkeley, and Jenner obtained some fluid from a sore on a dairymaid's hand and applied it to a half-inch, superficial incision of the skin of the arm of James Phipps, an eight-year-old healthy lad. The inoculation succeeded, and the boy developed a small pustule followed by scabs and a small scar. Six weeks later, Jenner boldly inoculated the boy with fluid from a smallpox victim. This dangerous step was crucial, since he was introducing small amounts of the live virus. If the vaccination with cowpox gave immunity, there should be no reaction.

James had no reaction. Jenner had urged some of his fellow physicians to try inoculations of pustule fluid as a means of preventing smallpox, but to no avail. He then followed this first experiment with another in which he took fluid from cowpox pustules and inoculated several others. These arm-to-arm transfers again resulted in no reactions. Jenner was now sure that the method was safe and effective. He wrote to a close friend, "I will now pursue my experiments with redoubled ardor."

Jenner sought to publish his findings in the *Transactions of the Royal Society*, but his manuscript was returned with the friendly admonition "that as he had gained some reputation by his earlier papers on the cuckoo bird's habits to the Royal Society, it was advisable not to present this, lest it should injure his established credit." Undaunted, Jenner was determined to publish it himself, and in 1798, he privately produced a small book with the imposing title of *An Inquiry into the Cause and Effects of Variolae Vaccinae, a Disease Discovered in Some of the Western Counties of England, Particularly Gloucestershire, and Known by the Name Cowpox*. This book is a milestone in the history of medicine, since for the first time, Jenner had developed a method to prevent rather than treat an infectious disease.

Of course, Jenner knew nothing about the immune system, but he did know that a weakened relative of the smallpox agent, which he labeled a virus, conferred protection against an infection by a disease-causing microbe. This was the path that Louis Pasteur would follow nearly one hundred years later, and one that would become the cornerstone of many future vaccines. Jenner quickly became so famous that the king allowed him to dedicate the second edition of his book to the king. By this time, the word *vaccinia*, the scientific name for the cowpox virus, had become part

of the medical language, and the inoculation with cowpox was called vaccination. Vaccination spread quickly and widely as people wanted assurance
that they would not be attacked by smallpox.

Opposition to Vaccination

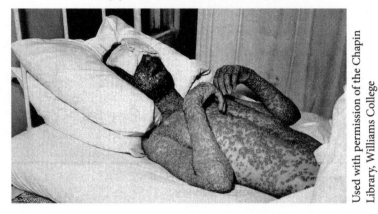

Used with permission of the Chapin
Library, Williams College

As might have been expected, disputes and criticisms soon broke out as
to the merits of inoculation with cowpox. Dr. Jan Ingenhousz, a notable
London physician, led intense criticism against the idea of cowpox con-
ferring immunity against smallpox. The criticisms from such a formidable physician weakened Jenner's position. Unbeknownst to Jenner,
another especially aggressive physician, a Dr. George Pearson, founded the
Institute for the Inoculation of the Vaccine-Pock with the duke of York as
its patron. Pearson's institution seriously deviated from Jenner's original
vaccination method and resulted in many serious outcomes. Hearing of
these events, Jenner went to London to rescue his discovery from destruction and to expose the many errors that had been committed by his critics.
He obtained the support of the duke of Clarence and Lord Egmont for the
establishment of a public institution devoted to providing free vaccination.
He also persuaded them to take the necessary steps to close Pearson's institution. Jenner's plan eventually became the Royal Jennerian Society for the
Termination of the Smallpox; Jenner became its first president.

But rashes of any sort that broke out over the bodies of their patients
were all blamed on vaccination. One physician went so far as to claim that
a year after being vaccinated, a child had developed an ox-face deformity!
Many frightened citizens visualized all sort of bovine facial deformities due
to the cow pock, which were captured by James Gillray's famous caricature,
The Cow Pock: The Wonderful Effects of the New Inoculation.

For the remaining years of his life, Jenner fought for vaccination against many physicians who condemned vaccination because of any remote "side effect." Jenner spent an inordinate amount of time in London, away from his practice and his family. Living in London for many weeks, he incurred very substantial debts.

Vaccination Spreads Worldwide

But despite Jenner's opponents, vaccination was catching on, and his reputation was growing not only in England but also on the Continent and in the United States. The influential Professor Benjamin Waterhouse of Harvard vaccinated his son, the first such inoculation in the United States. Waterhouse was so impressed with vaccination that he sought to achieve a monopoly on its use. He set his fee at five dollars. The monopoly lasted a month, since many Boston physicians condemned vaccination because it prevented them from collecting fees for the care of smallpox patients. Waterhouse persuaded President Thomas Jefferson to vaccinate his family and several hundred of his neighbors, an act that provided great prestige and support for the new preventative of smallpox.

Impressed by the difficulty of transporting the vaccine, Jefferson designed a two-part vial, consisting of an inner chamber containing the vaccine matter and an outer chamber filled with cool water to help prevent overheating. Vaccination spread widely thanks to the presidential blessing. Jefferson arranged for the vaccine to be provided to some Native American tribes. The Plains Indians were suffering from a smallpox epidemic that spread from the Gulf of Mexico to the Dakotas. Jefferson later wrote to Jenner assuring him that "mankind can never forget that you have lived."

In France, Napoleon ordered all of his soldiers to be vaccinated, and the empress of Russia urged all her subjects to be vaccinated. Leading physicians in all European countries hailed vaccination as a breakthrough against smallpox. Jenner's fame was worldwide. In 1808, the British government formed the National Vaccine Establishment under his direction. Five years later, during the war with France, Captain Milmari, a relative of Jenner's, was captured by the French and imprisoned at Verdun. Jenner wrote a letter to Napoleon requesting his release. On reading the letter, Napoleon exclaimed, *"Ah! C'est Jenner, je ne puis rien refuser a Jenner."* (Ah! I can never refuse Jenner.)

Later Days

Although he had achieved great fame, Jenner, now greatly in debt, sought a simpler life in Berkeley. His international fame brought gifts that helped relieve his financial plight. The English Parliament in 1802 granted him £10,000, and in 1807 an additional £20,000 was granted. Jenner returned to Berkeley and his old life in the country, treating patients, vaccinating everyone who wanted it, and enjoying the company of his friends.

But when his son Edward died of tuberculosis, Jenner fell into a deep depression. He began experiencing hallucinations, and in his depression, he withdrew from all of the activities that he had so enjoyed earlier in his life. Catherine, his wife, contracted tuberculosis; her death plunged him even deeper into depression. He developed seizures, and in 1823 he died during a severe convulsion. He is buried in the churchyard in his native Berkeley. James Phipps, the first person ever to be vaccinated, was one of the many mourners at the funeral service.

One hundred twenty-five years after Jenner's death, Sir Lionel Whitby, in his presidential address of 1948 to the British Medical Society, wrote the greatest posthumous honor ever paid to Jenner: "The end of the eighteenth century was marked by a special epoch-making adventure in experimental medicine, which was a prophetic prelude to the triumphs of the nineteenth and twentieth centuries, and which even today stands out as an unprecedented achievement in preventative medicine: Edward Jenner's successful cross-immunization experiments with cowpox."

Status of the Smallpox Virus and Vaccine

In 1980, the world began to breathe easier because of the successful World Health Organization (WHO) campaign to wipe out smallpox. WHO reported that the only stockpiles of the virus were in freezers that were located in the US Centers for Disease Control. Russia was known to have a stockpile of the virus, but was assumed to have destroyed it. But in the late 1980s, a Soviet scientist defector to Britain spoke authoritatively of active work being pursued by the Soviet Union. Margaret Thatcher, the British prime minister, angrily confronted Mikhail Gorbachev, the Soviet leader, to allow the world to learn the true state of affairs. Gorbachev invited a British team to inspect Vector, the Soviets' huge virology research complex in Siberia. During the visit the director, Lev Sandakhchiev, vehemently denied that the facility had any smallpox virus, despite the freely given response of a technician that he was indeed working on smallpox.

In 1992, Ken Alibek (formerly Kanatjan Alibekov), a leading Soviet bioweapons expert, defected to the United States. In his book entitled *Biohazard*, Alibek testified that there were twenty tons of liquid smallpox viruses still on hand at various Soviet military bases that could readily be loaded on missiles aimed at various American cities. In November 2002, the CIA, based on substantial evidence, reported that four nations, Iraq, North

Korea, Russia, and France, have secret stores of smallpox virus. Since then, suspicions have grown about the possibilities that rogue nations may also have smallpox capabilities for terrorist use. The CIA also believes that Al Qaeda may be acquiring biological weapons, perhaps including smallpox.

Many adult Americans who had been vaccinated as children are probably susceptible to smallpox because the initial response wears off with the time. In December 2002, the official US policy for those who are potentially at risk was issued. Smallpox vaccinations are recommended for smallpox response teams comprised of public health staff and healthcare workers likely to be involved in the initial care of any patients with smallpox. Smallpox vaccinations also are being offered to other healthcare workers and to first responders (including police officers, firefighters, and emergency medical technicians). Under current circumstances, with limited availability of licensed vaccine supplies, vaccination of the general population is not recommended, because the potential benefits of vaccination do not outweigh the risk of vaccine complications.

But many people in the second and third categories of voluntary respondents have declined to be vaccinated because of possible side effects and because epidemics seem so improbable. Furthermore, a significant percentage of the US population experienced some side effects to the vaccine, especially those with weakened immune systems and those with various skin diseases, most notably eczema.

A Hypothetical Epidemic

In the event of a terrorist attack with smallpox, the spread of the disease in reaching epidemic proportions would be frightening. Professor D. A. Henderson, an internationally acclaimed authority, has proposed the following scenario about how an epidemic of smallpox could occur. In what follows, it is important to remember that smallpox is an extremely contagious disease that is readily transmitted through air from victim to bystander. It is also the case that most adult Americans are not immune to smallpox, since their childhood vaccinations have lost their potency with time. In addition, one-third of the American population has never been vaccinated.

For the epidemic to be successful, two starting assumptions are needed. First, a bioterrorist attack of the smallpox virus would hit a target city having two hundred thousand or more residents. Among these people, some would likely be traveling to another sizable city or foreign country within the first two weeks of the attack. The other assumption is that the initial attack would reach one hundred persons. After infection, it would take two weeks for the disease to be recognized, since the ten-day incubation period is largely asymptomatic and since modern physicians are not trained to diagnose smallpox quickly, because it is so uncommon.

This is the first wave. Meanwhile, each of these hundred victims are transmitting the disease to ten others, creating a thousand new cases that form the second wave. If no one in this city traveled outside the city, the fourteen-day life cycle of the virus has no new outside victims, and the epidemic would be confined to that city and would eventually cease. The total number of victims depends on how rapidly emergency vaccination programs would be instituted.

But it is much more likely that some of these infected people would travel to another sizable city for business or pleasure during the fourteen-day period before the disease has been accurately recognized. But by then, it would be too late to vaccinate most of the first wave victims, and they would die of the disease. Once in the second city, the newly arrived infected asymptomatic persons would transmit virus to several thousand persons over the next week or two. These individuals in turn would perpetuate the transmission, creating another wave, which could easily grow to twenty thousand and keep on growing as this wave infected other vulnerable residents.

At this point, the transmission is out of control and the only prudent public health action is to promptly vaccinate everyone in both cities. Of course, if some of the later victims travel to another sizable city within the two-week period, another wave of victims would occur in that city. Were this process to be repeated a number of times, the situation would call for vaccinating a substantial fraction of the entire American population. Henderson estimates that one hundred million doses of the vaccine would be needed to stop such a surging outbreak that was originally triggered by one hundred initial cases of smallpox.

The means that the device used in a terrorist attack could be no larger than a credit card, a device that, when loaded with a small amount of concentrated smallpox virus, can emit an extremely fine spray of invisible microscopic droplets into the air that can linger for hours and can drift over a very wide range, such as a large airport terminal. In this case, the initial target cases would no longer number a hundred, as in the previous scenario, but would be several thousand, which would create an even more frightening picture.

A Better Smallpox Vaccine

A major breakthrough occurred with the announcement from Danish company Bavarian Nordic A/S of the development of a new and safer vaccine. This vaccine is made from vaccinia Ankara, a strain of the virus that differs considerably from the cowpox strain. According to Anders

Hedegaard, the CEO of Bavarian Nordic, the Ankara strain is a virus that has lost its genes that would have allowed it to replicate in the body, but has retained the genes needed to induce an immune response.

The company conducted trials of the new vaccine involving hundreds of subjects, many of whom had eczema or HIV, but none showed any of the side effects seen with the older vaccine. These trials and more are in progress. They showed that the vaccine elicits a faster immune response that the previous vaccine and full protection against lethal challenges in mice and in monkeys. It even provides protection when administered after the infection. Since this vaccine is injected subcutaneously or intramuscularly, it leaves no scar. Based on the trials showing that the new vaccine elicited the same immune response as the previous one, the FDA approved the vaccine for stockpiling, which is in progress with the goal of stocking three hundred million doses.

Comment

Many years earlier, Edward Jenner had accidentally overheard a conversation in which he learned that a cowpox infection on the hands of milkmaids conferred protection against the future development of smallpox. Subsequently, the folklore provided abundant confirmation in England and in Europe. But even though he did not know how that was accomplished, he concluded that the causative agent, which he called a virus, was transferable, since the only transmission to the milkmaids had to be the virus carried from the teats of the cow to the milkmaids' hands. Once again, we see a prepared mind at work.

REFERENCES

Barquet, N. et al. "Smallpox: the triumph over the most terrible of the ministers of death." *Ann Intern Med.* 127 (1997), 635–642.

Friedman, M., and G. W. Friedland. "Edward Jenner and Vaccination." Chap. 4 in *Medicine's 10 Greatest Discoveries.* New Haven, CT: Yale University Press, 1998.

Henderson, D. A. et al. "Working Group on Civilian Biodefense. Smallpox as a biological weapon: medical and public health management." *JAMA* 22 (1999), 2127–2137.

World Health Organization. Global Commission for Certification of Smallpox Eradication. *The Global Eradication of Smallpox: Final Report of the Global Commission for the Certification of Smallpox Eradication.* Geneva: World Health Organization, 1980.

10

Penicillin: The Crucial Role of Weather

The phrase that heralds a new discovery is not "Eureka!"
It's "That's funny."

—Isaac Asimov, author

Introduction

The story of penicillin has been told many times, often superficially and nearly always inaccurately. The following is a more complete account of this amazing discovery by Alexander Fleming. It deals with the critical role that the London weather played during September 1928. The unsung hero of this tale is Howard Florey, leader of the Oxford Group, whose great contribution was to bring penicillin from the laboratory to the bedside.

The Discovery

In 1928, Dr. Alexander Fleming, nicknamed "Flem," was happily working in a cramped lab in St. Mary's Hospital in London. Fleming's job was to isolate the bacteria that were causing diseases in patients. To do this, he would take a small, circular, covered glass plate called a Petri plate that was filled with nutrients on which bacteria would grow. Fleming would quickly lift the plate's lid and smear a sample of some infectious fluid like pus or throat washings that had come from one of the patients. Speed was essential to minimize any contaminating bacteria or molds in the air from accidently dropping onto the plate when it was exposed. He would then put the plate into a warm (37°C) incubator to speed the growth of the bacteria. The next day the bacterial growth would have spread all over the plate. Fleming would then identify the bacteria under a microscope and determine to which bacterial family it belonged. He sent his reports back to the doctors in the hospital, who depended on them to guide the treatments of their patients.

Fleming was actually quite good at what he did. As the warm weather of August 1928 rolled in, Fleming left the hospital for his holiday. Before leaving, he left several dozen plates on his lab bench that contained various strains of staphylococci bacteria to see if they were stable over a longer term at room temperature. He was following a method for such experiments that had been published by Professor Joseph Bigger.

Fleming wanted the results of this experiment to use in a lengthy chapter that he was writing on staphylococci for an important multivolume treatise with the title of *A System of Bacteriology in Relation to Medicine*. While he was on his holiday, the bacteria on those plates grew slowly, since the temperature in the laboratory was about 20°C, considerably lower than the incubator.

Fleming returned to his lab after his five-week holiday. Merlin Pryce, an old colleague of Fleming's, dropped by to say welcome back. Fleming was busy examining the old plates he had left on the bench. He examined each plate and recorded the growth of bacteria on it in his notebook. He then stacked the plates and moved them to a shallow Lysol bath that was used for decontamination. As the plates in the stack were slowly sinking in the bath, Pryce noticed that the top one, still above the level of the bath, contained a white fluffy mold, around which was a clear zone where the bacteria had been killed.

Pryce pulled this plate out just as it was sinking in the disinfectant bath and gave it to Fleming, who looked at it and said, "That's funny." They both had seen molds before that accidentally dropped onto plates from the air at the time when the bacteria samples were first being added or when the lids were quickly removed and replaced in order to permit a better view of the bacterial colonies. Always before, such mold-contaminated plates were discarded. But on this plate a mold had grown, and it seemed to kill the bacterial growth on the plate, a defined space adjacent to the mold but not in those areas of the plate that were more distant from the mold. This was something new! Fleming described the discovery in his own words in his 1929 paper:

While working with staphylococcus variants, a number of culture plates were set aside on the laboratory bench and examined from time to time. In the examinations, these plates were necessarily exposed to the air and they became contaminated with various microorganisms. It was noticed that around a large colony of contaminating mold, the staphylococcus colonies became transparent and were obviously undergoing lysis [i.e., breakdown]. Subcultures of the mold were made and experiments carried out to ascertain something of the properties of the bacteriolytic [i.e., bacterial killing] substance, which had evidently been formed by the mold culture and which had diffused into the surrounding medium.

Fleming took a sample of the mold and grew it in a nutrient broth. The broth turned yellow, and when samples of the broth were tested, they killed many different types of bacteria. He also took a sample to the mold lab, and it was identified as a rare and unusual member of the common bread mold family, penicillium.[2] It seemed very likely that a mold spore of the penicillium family had drifted up the staircase from the mold lab, which was located one floor below Fleming's lab, and into Fleming's lab, where it landed on one of the plates on the bench.

One of the mysteries concerning Fleming's discovery was why he waited six weeks before he recorded his first observation of this unusual event in his lab notebook. Robert Root-Bernstein, a distinguished medical scholar, has proposed that when Fleming first saw the plate with the mold appearing to have killed the bacteria, he thought that it was due to the mold secreting lysozyme. Fleming had a seven-year research interest in lysozyme and had published eight papers reporting his research.[3]

In one experiment, he did show that this newly discovered mold juice substance (i.e., the broth) was very active against several bacterial strains in which lysozyme was inactive. Fleming continued to keep the unusual mold growing in the broth and used it as a playful tool. For example, he would streak a plate with various bacterial colonies of different colors, and then using drops of the liquid broth, he would create colorful designs. Fortunately, Fleming did publish a brief paper describing his results in a

---

2 The term *penicillium* was derived from Latin *penicillium*, or resembling a painter's brush. Fleming thought that the term *penicillin* sounded better than "mold juice," which was used during the preliminary studies.

3 Lysozyme (Greek *lysein* + *en*, within in, and *zyme*, ferment) is an enzyme with antibacterial properties. It is normally present in saliva, sweat, breast milk, and tears. It has no medical uses.

1929 issue of the *British Journal of Experimental Pathology*. It attracted very little attention.

Even though Fleming had been trained as a doctor, he didn't foresee that the substance in the mold juice might be useful in treating infections in patients, nor did any of his colleagues. Fleming's boss, a crotchety old gentleman named Sir Almroth Wright, had had a brilliant career as a young man in developing vaccines. But at the time, very few people in medicine could ever imagine that a substance of microbial origin could cure an infection.

Howard Florey and Ernst Chain

Fleming's paper lay on dusty library shelves for ten years until two scientists named Howard Florey and Ernst Chain, working at Oxford University, discovered it and saw its potential application. Florey, a professor at the William Dunn School of Pathology at Oxford, had recruited Chain, a talented twenty-nine-year-old chemist, originally from Nazi Germany, who had just completed his PhD in biochemistry at Cambridge. Florey had been working on lysozyme and of course knew of Fleming's work in that field. He dispatched Chain to pursue a library search for other natural occurring substances that inhibit bacterial growth.

Chain finished a long afternoon of reading, and when he left the library, he ran into a technician in the hallway just outside the library. Chain was a friendly, charming man, and he stopped to chat with Miss Margaret Campbell-Benton, who had worked with the now deceased Professor George Dryer. Chain noticed that Miss Campbell-Benton was carrying a test tube containing a yellow fluid. She said that it contained a culture of the fluid *Penicillium notatum*, a mold that Fleming had sent several years before to Dryer for his unsuccessful search for viruses that infect bacteria. Nevertheless, as was the routine practice in the laboratory, she had kept the mold culture growing as part of a large reference group of cultures that were being maintained for possible future use.

She told Chain a little about Fleming's discovery of the killing power of the yellow mold juice, adding that he had described it in a 1929 paper. Chain's mind clicked! He rushed back into the library, read Fleming's paper, and hurried to tell Florey that this was the project they should consider taking on. Both Florey and Chain thought that the mold might be producing lysozyme, but it was easy to find out. Perhaps it was something else.

Florey was impressed with this newly uncovered paper, and he and Chain launched studies aimed at discovering the mechanism of its action on bacteria. It soon became clear that the substance being secreted by the mold was not lysozyme, but something else of considerable and wider potency. Chain, with his chemical expertise, took on the problem of how to separate penicillin from all the other substances in the yellow broth. He worked diligently but was unsuccessful, since each time he tried a new approach, the penicillin activity disappeared. He finally realized that he would have to try to separate penicillin at a cold temperature rather than at room temperature.

Chain knew and admired Norman Heatley as an innovative and skilled laboratory assistant, and Chain hired him to help with the problem of keeping the entire process in a cold temperature. Heatley, also a talented laboratory gadgeteer, built a "Rube Goldberg" apparatus that kept all the steps of the purification process at zero degrees. It worked, and Chain produced enough penicillin to try out on a few animals.

The Animal Experiments

Florey personally took over the animal experiments. He injected eight white rats with enough streptococci to kill them in twenty-four hours. Four received penicillin and four did not. The team set up a round-the-clock vigil. At 10:30 p.m., Florey saw that the untreated animals looked sick but the other four seemed normal. At 3:30 a.m., Heatley found that the four untreated rats were dead, but the four penicillin-treated animals were as lively as ever. The reader can imagine the excitement of the laboratory team the next morning when they heard the great news! More animal experiments using lethal doses of different bacteria confirmed the original results. Penicillin was on its way to becoming a "wonder drug."

With these animal results in hand, Florey was determined to get enough penicillin to start treating a patient. He converted all of the available laboratory space to growing the mold. Since it took many liters of the broth to yield a single dose of penicillin, every type of shallow vessel was called into service, including pie pans, cookie tins, a dog's bathtub, and a large number of hospital bedpans. All of these were shallow vessels that allowed the mold to grow on the surface, while minimizing the volume of mold juice beneath.

This was at the time when Britain was on her knees due to the German bombings, and the belief that England would be invaded by German troops was widespread. Fearing such an invasion, Florey's team put penicillin mold spores in their clothing so that any survivor of the group would be able to carry on the work.

Treating Patients

The first patient that Florey treated was a local policeman who had severe staphylococcal and streptococcal infections from the scratch of a rose thorn. The man was near death. Over the first three days of treatment, he improved markedly, but the limited supply of penicillin was running dangerously low. Florey knew that some of the injected penicillin was excreted in the urine unchanged, and he ordered the policeman's urine to be collected and quickly brought from the hospital to the laboratory, where it was recycled for its penicillin content.

Florey's wife, Dr. Ethyl Florey, organized volunteers for what onlookers called the "P-patrol." The policeman was treated for two more days until the supplies of recycled penicillin were completely exhausted. The fever and infections recurred, and the man died. Later a wag remarked, "Penicillin is a substance grown in bedpans and purified by passage through the Oxford Police Force!"

Florey contacted several British drug companies to see if they would produce larger amounts of penicillin. All of his requests were denied, since they were running at full capacity to meet pressing wartime needs. In early 1941, Florey flew to the United States and tried to persuade several American drug companies to take on the challenge. He did get a commitment for a limited supply from one company for penicillin to be delivered later in the year. But more importantly, a large laboratory of the US Agricultural Department in Peoria, Illinois, had developed techniques that made large-scale production possible.

In December 1941, the United States was plunged into the war, and all production of penicillin in America was now redirected for its own use. Back in England, Florey finally persuaded the Imperial Chemical Industries to produce enough of the drug for a large-scale trial, and over the next year, 172 patients were treated with spectacular results.

On August 5, 1942, Florey received an urgent telephone call from Fleming, pleading for some penicillin to treat fifty-two-old Harvey Lambert, a close friend of Fleming's who had streptococcal meningitis, which was regarded as a uniformly fatal disease. Fleming had performed a spinal tap and had cultured the spinal fluid, which was full of teeming streptococci. In his laboratory Fleming had found that the streptococci were resistant to added sulfa drugs but were readily killed in a test tube when some crude penicillin soup was added.

Although the Oxford patient trials had severely limited his supply, Florey felt he could not refuse the discoverer of penicillin. He boarded the next train and hand-delivered the penicillin to Fleming at St. Mary's Hospital. He stayed long enough to instruct Fleming on the details of administration and then returned to Oxford. The penicillin was administered at two-hour intervals intramuscularly to Mr. Lambert for five days with little clinical improvement.

Fleming wondered if the drug was penetrating the brain and reaching the infected sites in the brain, since many drugs do not cross the blood-brain barrier. Perhaps it ought to be given directly into the spinal canal. Fleming called Florey for advice about this idea, but Florey was reluctant to approve such an untested procedure, although he did agree to test it in an animal. Hanging up the phone, Florey immediately went into his laboratory and injected penicillin into the spinal canal of a rabbit, which died almost immediately. He rushed to phone Fleming to tell him not to proceed, but was astounded to hear that Fleming had gone ahead and given the drug in the spinal canal. Not only did his friend survive, he made an amazing and complete recovery.

This miraculous cure led the London *Times* to publish a feature story on penicillin. Fleming became an instant hero. The press continued to hound him for more information, and he came to enjoy the glory that was being heaped on him as a celebrity. Florey and his team, however, decided to stay out of any limelight and avoided the press completely. It is a bit ironic that even though the Oxford Group had successfully treated 172 patients, Fleming had treated only one.

Thanks to the large-scale production of penicillin in Peoria, Illinois, it soon became known far and wide as a wonder drug. The success of the drug spread first to the battle-wounded troops and then to a wide variety of infections in the civilian population. Today there are twelve chemical variations of the basic penicillin, and each has special uses for specific diseases.

As the medical successes due to penicillin piled up, so did the media's praise of Fleming. He certainly was a willing partner with the press, always enjoying interviews and stories about him and his discovery. Lord Moran, later Winston Churchill's doctor, was the head of St. Mary's Hospital and had valuable connections with Lord Beaverbrook's newspapers.

Fleming's fame grew as Moran increasingly assumed to be his unofficial press agent and for raising funds for St. Mary's. Many of the articles put the success of the Oxford efforts on Fleming's shoulders, much to the distress of Florey, who continued to ignore the press but at the same time was trying to correct their mistakes. Fleming was aware of these errors but did little or nothing to correct them.

In 1942, the London *Times* ran an editorial praising the success of penicillin without naming any of the scientists involved. In response, Sir Almroth Wright, Fleming's boss, published a letter, an excerpt from which follows: "In your leading article of yesterday on penicillin, you refrained from putting the laurel wreath for this discovery round anybody's brow. With your permission, I would supplement your article by pointing out that on the principle of *palmam qui meruit ferat* [honor to one who earns it], it should be decreed to Professor Alexander Fleming of this research laboratory."

The Nobel Prize

The nominations for the 1945 Nobel Prize in Physiology or Medicine were fraught with controversy. Fleming's name was the most often promoted as the best candidate by acknowledging his discovery and by his great support from the British and American popular press. On the other hand, Florey and Chain were supported by the more enlightened public and scientific press of both countries. But even *The New York Times* ran a headline that listed Fleming and "two coworkers." Professor John Fulton of Yale, a close friend of Florey's dating back to their student days at Oxford, telegraphed the paper asking that they immediately correct their egregious error of omitting Florey and Chain. This monumental error was corrected the next day.

In 1945, the Nobel Prize committee voted to award the Prize in Physiology or Medicine to Fleming, Chain, and Florey. But no mention was made of Norman Heatley, despite the fact that those who knew the inner workings of the team had long concluded that "without Heatley, there would have been no penicillin." Later Heatley was honored by Oxford when he received the first honorary degree of Doctor of Medicine in its eight-hundred-year history.

The Aftermath

Alexander Fleming died a quiet death. As his last request, he asked his wife to comb his hair so that "I am decent." Fleming was buried in St. Paul's, a high honor reserved for only a few illustrious Englishmen. A flagstone bearing the inscription of "A. F." marks the site of the crypt. On a wall nearby is a marble tablet, on which the thistle of Scotland and the lily of St. Mary's Hospital are carved, the two places that Fleming loved.

In 1968, Florey died of a massive heart attack at the age of sixty-nine. He is buried in Westminster Abbey. At the memorial service, Lord Adrian, himself a Nobel laureate, said of him, "He was one of the great leaders of medical science. Millions owe their very lives to him, and to what he did . . . Florey is to be honored just as were Pasteur and Jenner and Lister."

Ernst Chain left Oxford in 1948 after a final dispute with Florey over patents and royalties and accepted a position as director of chemical research at the Institute of Health in Rome. He died in 1979 at the age of seventy-three.

A Mystery Solved

Although Fleming certainly discovered penicillin, one of the great mysteries was why he was never able to repeat his original finding. Many years later, Professor Ronald Hare discovered that the change in the weather of London during the period of Fleming's holiday in 1928 was the answer. Hare obtained detailed data from the London Weather Bureau, charted the daily highs and lows over the period of Fleming's holiday, and discovered a pattern of an initially cool spell followed by an unusually warm spell.

Hare reasoned that when the bacteria were first plated and the mold accidentally had dropped in as a contaminant, the lower temperature favored the growth of the mold, giving it a kind of "head start" over the bacteria, which grew faster during the period of the higher temperature, hence creating the pattern found by Fleming. Hare also showed that when the mold and bacteria were both planted and subjected to a warm-warm sequence, no inhibition of bacterial growth by the mold occurred. The mystery was solved by Hare's studies.

But fortunately Fleming, unaware at the time of these temperature effects, saved a sample of the mold and subcultured it so it was available for future studies by him and by others. Had he not done so, we would not have any penicillin today.

The Role of Chance

The role of chance in Fleming's discovery has been likened to the chances of winning the following seven horse races in a precise sequence:

* Chain's chance conversation with Margaret Campbell-Benton led to his obtaining a subculture of Fleming's mold and her direction to consult Fleming's 1929 paper.
* The plate was planted with staphylococci that happened to be sensitive to penicillin; had a strain of staphylococci that was resistant to penicillin been used, nothing would have happened.
* The mold contaminant that dropped into the plate was a rare variety of the penicillium family that produced penicillin and came up from the lab immediately below.
* The plate was left on the lab bench, subject to the swings of the outside temperature.
* The initial cool spell followed by the warm spell changed the lab temperature in the right sequence to favor mold growth first followed by bacterial growth.
* The plate was placed on top of a stack of other plates in a Lysol bath, but Pryce happened to grab the plate and presented it to Fleming, who uttered, "That's funny!"

* Instead of discarding the plate as a common contaminant, Fleming saved a sample of the mold, which he studied as a playful tool, never dreaming of its possible use in infectious diseases.

REFERENCES

Hare, R. *The Birth of Penicillin & the Disarming of Microbes*. London: Allen & Unwin, 1970.

Hobby, G. *Penicillin: Meeting the Challenge*. New Haven: Yale University Press, 1985.

Lax, E. *The Mold in Dr. Florey's Coat: The Story of the Penicillin Miracle*. New York: Henry Holt, 2004.

MacFarlane, G. *Alexander Fleming: The Man and the Myth*. Oxford: Oxford University Press, 1984.

Streptomycin: A Major Dispute

Al has hit pay dirt.
            —*Doris Jones, graduate student*

Introduction

In early 1947, a doctor made the following notes in the chart of a twenty-one-year-old woman who had an advanced stage of tuberculosis of the lung. Treatment had begun on November 20, 1944, and lasted four months, during which she received five courses of a new treatment. She improved markedly, and two years later, her case was closed as arrested pulmonary tuberculosis. She subsequently married and had three children.

Streptomycin was the drug that this young woman received, and she was the one of the very first patients to benefit from this miraculous drug. Many hundreds of thousands of other tuberculosis patients followed. These grateful patients may not have realized that a graduate student studying under a microbiologist named Selman Waksman, working at Rutgers University in New Jersey, had discovered streptomycin, which was the miracle drug that had arrested her disease. Waksman was a deliberate, systematic scientist who spent his professional life studying the microbes in soil. Waksman provided close supervision of the work of several graduate students, one of whom, Albert Schatz, would discover this life-giving medicine.

Waksman was the son of Jewish parents born in 1888 in the village of Novaya Priluka in Ukraine. Because of its prevalent anti-Semitism, the seventeen-year-old Waksman left Czarist Russia to follow his dreams in the United States. In 1915, as a student at Rutgers University, he became deeply interested in the study of a group of soil molds called the actinomyces. Two years later, he published his first paper on a particular mold species known as *Actinomyces griseus*.

Little did he realize that forty years later, this same mold would yield streptomycin! As a young faculty member at Rutgers, Waksman was committed to a study of molds in soil. For Waksman, soil was not a mass of decomposed plant and animal remains, but rather a site teeming with bacteria, molds, protozoa, and other organisms. Waksman knew that many disease-causing organisms of humans, such as typhoid bacillus, disappeared quickly when they were deposited in soil.

He reasoned that this must be due to the activity of soil microbes that are antagonistic to the pathogens, a process he named antibiosis. But throughout the 1920s and 1930s, Waksman's work was tightly focused on other aspects of the many soil molds. His numerous papers and several books of that period hardly mention any antagonism of one microbe against another.

Missed Opportunities?

Waksman had several early opportunities to switch his focus to studying certain soil microbes that might kill tuberculosis bacteria, but he failed to follow them up. In 1932, he received a government grant designed to learn why the tubercle bacillus, once it had been deposited in soil, disappeared. But his work under the grant concentrated on a distantly related problem, and the chance to discover possible antibiotics against the organism was missed. In 1935, Waksman was presented with a second, even better opportunity. Dr. Fred Beaudette, a pathologist at the New Jersey Agricultural Experiment Station at Rutgers, showed Waksman a culture plate of tubercle bacilli that had been killed by the chance contamination of a mold.

This was reminiscent of Alexander Fleming's discovery of penicillin in 1929, but at that point, Waksman did not realize the potential import of such an observation. He still preferred to pursue his basic studies of soil microbiology. He may have been dissuaded by the then-prevalent view that killing the tubercle bacillus by chemicals was hopeless. Many had been tried in the past, always unsuccessfully. The organism, unlike other bacteria, had a fibrous, waxy, external coat that was thought to retard penetration of any drug.

Waksman's son, Byron, who was a medical student at the University of Pennsylvania, presented his father yet a third opportunity by urging him to take up the search for an anti-tuberculosis agent. Byron was quite familiar with his father's work since he had read some of the papers that the elder Waksman had published. Young Waksman wrote the following letter with a proposal for a summer project: "In reading the reprints you sent me . . . I was particularly impressed with the relative simplicity of the method you have used in isolating fungi-producing antibiotic substances, and I wondered if exactly the same method could not be used with equal ease to isolate a number of strains of fungi of Actinomyces which would be effective against M. tuberculosis. There is no question that it has a great deal of practical value." The senior Waksman's response to the letter was "the time has not come yet." But the time was fast approaching.

The Discovery of Streptomycin

Waksman's best graduate student seeking a PhD degree was a twenty-three-year-old student named Albert Schatz, who undertook a systematic screening of several thousand samples of soil organisms looking for one that might kill various pathogenic organisms, especially any that were resistant to penicillin. It was a formidable task, since there were many variants of every class of soil molds. Working day and night for several months, Schatz was obsessed by his search, since obtaining his PhD degree was dependent upon submitting a thesis that contained new positive results. Negative results, even those that were the results of an intense search, weren't acceptable thesis material. But after slaving for several months in testing several thousand samples from the soil, Schatz came across a mold from the large actinomyces family that was positive in his test system. Curiously, the culture of this mold originally had come from a throat swab of a sick chicken and not from the soil. Schatz had obtained the culture by chance because he had requested that his fellow graduate students give him their cultures before discarding them in the hope of using those cultures to widen his search from sources other than the soil.

The culture from the sick chicken had come from Doris Jones, who had passed the plate though a window from her lab to Schatz's lab. The mold proved to be a powerful killer (later named streptomycin) of the strains of penicillin-resistant, pathogenic bacteria that Schatz was looking for. Moreover, this mold exuded streptomycin, which killed the tubercle bacillus. Albert Schatz and a fellow student, Elizabeth Bugie, along with Waksman reported this discovery in 1944 in a three-page article. The order of authors on the paper was Schatz, Bugie, and Waksman. (Elizabeth Bugie was a graduate student who worked briefly with Schatz, but was assigned to other unrelated projects. She never claimed any role in the actual discovery.) In the publication of such papers in the sciences, the senior author is nearly always the last to be listed, as it was in this case. But this particular sequence of authors was to surface later as a major sticking point.

The paper contained a single table in which streptomycin activity was tested on the tubercle bacillus as one of nineteen different bacterial species that were listed halfway down in a column. It had killing power toward the tubercle bacillus as well as toward some of the other species, but the tubercle germ was not singled out. Curiously, the potential significance of this important finding was not mentioned anywhere in the paper. It seemed that a great discovery had been made, but the discoverers had not as yet discovered it! At this point, the reader must wonder how these scientists could have possibly overlooked such a dramatic finding.

Feldman and Hinshaw

William Feldman was a veterinary pathologist who had joined the staff of the Mayo Clinic in the department of comparative pathology. Corwin Hinshaw was a physician in the Mayo department of pulmonary diseases. Having complementary skills, the two were ideal collaborators. Furthermore, they were persistent in their search for an effective drug to treat tuberculosis. It was a heady task. By 1932, over a thousand papers had been published in which no fewer than thirty different agents had been tried, all unsuccessfully. Hinshaw and Feldman had been conducting trials on new derivatives of the various derivatives from the mid-1930s onward, but none had worked in curing the disease. In fact, during that last few months of 1943, they were making preparations for a large-scale trial of yet another drug derived from the sulfa-drug family.

Feldman's professional life seemed to be totally and energetically dedicated to tuberculosis and its treatment. In his enthusiasm, which may have been influenced by the early successes of penicillin, he had heard that Selman Waksman's laboratory at Rutgers University in New Jersey was searching for a new antibiotic, and that Waksman had had a few meetings with others about looking for something like penicillin for the tubercle bacillus. In his zeal, Feldman made a quick visit to Waksman. At the meeting, Feldman strongly urged Waksman to always include the tubercle bacillus as a target in future searches. At that time, Feldman was completely unaware of the work of Albert Schatz in his basement laboratory that had led to the discovery of streptomycin only five weeks earlier. Waksman declined to reveal that fact to his visitor.

During the discussions, Feldman told Waksman that in their searches, they should use a highly virulent strain of the tubercle bacillus but always insisting on strict sterile conditions during its handling to avoid infecting anyone. He cautioned that unless they used this highly virulent strain, they could easily be led astray. Although Waksman had fears of allowing such a potentially dangerous organism in his laboratories, he requested that Feldman send him cultures. Later, Schatz volunteered to be the only worker to handle all of the procedures that involved the highly virulent strain. In leaving, Feldman felt that he had persuaded Waksman of the expertise and enthusiasm of Hinshaw and himself in seeking new measures to treat what was called the Great White Plague.

The Schatz, Bugie, and Waksman paper appeared in print in January 1944. Later, Waksman revealed that the tuberculosis strain in the paper was a non-virulent type. But that was not important when Schatz later showed that the virulent strain was even more vulnerable to streptomycin. It is a curious lapse that neither Waksman nor his coauthors saw the implication of a drug that killed the tubercle bacillus. Over the next few months, Schatz produced enough streptomycin to supply a well-controlled trial in animals. Waksman then contacted Feldman, indicating that he now had enough streptomycin for the trial. In April 1944, with the streptomycin supply in hand, Feldman and Hinshaw divided forty-eight guinea pigs, half into a control group and the other half into a treated group.

All animals in both groups were given fatal doses of tubercle bacilli. The treated group received streptomycin. The results were decisive. Only the control animals developed the disease and died. The treated animals showed little or no sign of tuberculosis infection. Scoring the extent of the infection, with 100 as the maximum, the score of the non-treated control animals was 82, while the streptomycin-treated group scored less than 5.

Treating Patients

These dramatic results established streptomycin as an urgent contender for human trials, and within nine months, Hinshaw and Feldman had successfully treated thirty-four patients. An avalanche of publicity greeted their announcement. Finally, it seemed that the Great White Plague, one of the most dreaded of diseases, had been conquered. Desperate patients and their families from all over the world sought the new miracle drug. An international streptomycin black market sprang up almost immediately. Streptomycin and the drugs that followed had a profound impact on the treatment of tuberculosis. Over the next four decades, a whole new approach to treatment with drugs was launched and in full swing. Mortality rates plunged to less than 1 percent. Tuberculosis patients, once consigned to sanatorium care and often to lung surgery, no longer needed such questionably effective measures. Sanatoriums began closing, decreasing from four hundred thousand to nearly zero in fifteen years. Patients reaped other benefits that had been denied them because of their disease.

Once they were considered unsafe marriage partners and risky employees. Drug treatment changed all of that; no longer did they carry the label of having "TB" or as being a "lunger."

Merck and Company, which had been instrumental in manufacturing small amounts of the drug for the early studies, now went into full production to meet a market demand that exploded overnight. Physicians were excited by the new drug's curative effects, especially in tuberculosis of the lungs (called "consumption"), tuberculosis meningitis, and the generalized infection known as "galloping consumption." There were some side effects, the most serious being damage to the part of the ear that affects hearing and balance. Furthermore, drug resistance began to develop, a problem that loomed larger and larger as time passed. Fortunately, another drug was being developed in Sweden that would help solve this problem.

Another Anti-Tuberculosis Drug

The route of the discovery of streptomycin had been the result of a painstaking screening that involved thousands of microbes. It was completely different from Fleming's accidental discovery of penicillin. The route of discovery of the new Swedish drug called para-amino-salicylate (or simply PAS) followed a third path. It was based on attacking the basic biochemistry of the tubercle bacillus. This discovery started with the curious observation that the metabolism of the tuberculosis organism was stimulated by sodium salicylate, a close relative of aspirin. It seemed that salicylate accelerated some essential metabolic process; it was in a sense a food stimulant for the bacteria.

At the same time, Oxford scientists found that a compound that was a "look-alike" to sulfanilamide, called para-amino-benzoic acid (or simply PABA), competed with sulfanilamide for uptake of the sulfa drug by some bacteria. This was the origin of the concept of competitive inhibition of one agent against another. Jorgen Lehmann, a biochemist in Gothenburg, adopted the Oxford strategy and screened a number of compounds that were "look-alikes" to sodium salicylate to see if its stimulating effect was diminished or abolished by any of the test compounds.

Lehmann found that PAS not only blocked sodium salicylate uptake, but it also produced a marked inhibition of the growth of the tubercle bacillus. Lehmann went on to conduct animal experiments and clinical trials at about the same time that similar trials of streptomycin were underway in the United States. The progress on streptomycin in the United States and on PAS in Sweden closely paralleled each other, and the respective publications occurred at nearly the same time.

But Hinshaw and Feldman published the results of their clinical trials with streptomycin ahead of those of Lehmann with PAS and thus claimed priority of discovery of streptomycin as the first proven anti-tuberculosis drug. PAS proved to be most effective when combined with streptomycin. Such "combined therapy" of the drugs complemented each other and reduced the occurrence of resistant organisms. Ironically, Dr. Feldman was one of the first to benefit from the combined therapy. He had contracted tuberculosis from his close contact with the diseased guinea pigs that had proven so valuable in the earlier studies. Thanks to the combined therapy, he had a complete recovery.

Albert Schatz's Claim

In 1949, Dr. Albert Schatz shocked the scientific community by filing a lawsuit against Waksman and Rutgers University. Schatz had enjoyed a cordial correspondence with his old professor over the six years following his receiving his PhD degree, but in January, he wrote to Waksman raising some embarrassing questions as to the fate of the royalty monies being generated by the sale of streptomycin. Waksman did not appreciate the tone of the letter, and he quickly responded that Schatz's contribution to the discovery of streptomycin was "only a small one."

Waksman had already received $350,000 from royalties ahead of all of the remaining royalties that were then paid to Rutgers University. At the time, the public was under the impression that all royalties from streptomycin were being donated to Rutgers, but in fact, Waksman continued to receive a significant fraction, perhaps as much as half of the total. Waksman had also sent Schatz $1,500, but Schatz, unaware of Waksman's newfound wealth until later, treated the money as a gift and declared it as income on his tax return, whereas Waksman wrote it off as pay for work done by Schatz.

Schatz, upset by the tone of Waksman's letter that demeaned his efforts in the discovery, filed a lawsuit in May 1949 in which he sought recognition as the co-discoverer of streptomycin and equal participation with Waksman in the monetary benefits. He had good reason to support his claim. His PhD thesis detailed much of the basic work on the discovery. Furthermore, he argued that he was the first author on the first paper in 1944 announcing the discovery, and that he was a coauthor on several other early papers. Finally, he pointed out that he had been listed as co-discoverer on the patent on streptomycin filed in 1945 and issued in 1948. This was probably his most potent argument.

News of the lawsuit was met with widespread outrage by many members of the scientific community, who found it difficult to believe that a lowly graduate student actively seeking his doctoral degree under a distinguished professor could lay any claim to any discovery. The "old boy network" of senior scientists and their apostles closed ranks in support of Waksman. The opinion of the orthodox scientific community was that although Schatz may have discovered the streptomycin-producing organism, it was really Waksman's nurturing scientific environment in which Schatz, a mere student, worked that made it possible.

Schatz asked for an accounting of the royalties that were rapidly accruing to the university from which Waksman was receiving half. He felt he was entitled to half as well. Although the first royalty check received by the university had amounted to a measly $73.10, by the time of the Schatz affair, royalties of several million dollars had accumulated and were growing rapidly. After eight months, the lawsuit was settled, and the royalties were redistributed. Waksman would receive 10 percent, Schatz 3 percent, and a group of others who had worked on the discovery, a total of 7 percent. The university retained the remainder. Waksman subsequently assigned half of his 10 percent share to build an Institute of Microbiology at Rutgers, subsequently renamed in his honor.

Another Lawsuit

In a separate lawsuit, a dermatologist named Dr. Mary Marcus claimed that she had given Waksman the cultures that twenty years later produced streptomycin. As the presumed inventor of streptomycin, she was therefore entitled to $5 million, subsequently raised to the "more reasonable [sic] figure" of $20 million! Marcus's attorneys soon found that they were dealing with an eccentric client. Over a two-year period, she refused to give any pretrial depositions or to assist in other ways in preparing the case for trial. Meanwhile Professor and Mrs. Waksman were being subjected to some uncomfortable and bizarre events. Letters to the Waksmans in passé Russian threatened attacks on Dr. Waksman's reputation if the suit wasn't settled. One letter went so far as to warn Waksman, a prominent Jew, that he could expect malicious attacks from "Jew-hating attorneys." The Waksmans endured many unpleasant days during the two years, which thankfully were ended when the judge dismissed the case.

The Nobel Prize

In 1951, the Nobel Prize committee was drawing up a list of possible contenders for the coveted prize in Physiology or Medicine. Waksman's name was high on the list, but since the rules governing the prize allowed for no more than three awardees, there was speculation about whether the committee would entertain any others. There were several possibilities: Albert Schatz was one, William Feldman and Corwin Hinshaw were two, and Jorgen Lehmann—with or without his coworkers Karl Rosdahl and Frederick Bernheim, for their discovery of PAS—were one or three.

The deliberations of the committee are always secret. But to gain information, they send delegates to visit and evaluate all serious candidates before they make any final decision. These representatives visited the Mayo Clinic to collect information about Feldman and Hinshaw. Lehmann also received such a visit. Since Waksman became the sole final winner, these other visits had evidently not been successful in persuading the committee of the merits of any of the contenders that were equal in a scientific sense to that of Waksman. In the case of Lehmann, the problem may have been the matter of dating the priority of his discovery in the treatment of patients with PAS, since his paper came later than the published success of streptomycin.

Waksman was the sole recipient of the 1952 Nobel Prize "for his discovery of streptomycin, the first antibiotic effective against tuberculosis." In his Nobel address, he never mentioned the names of Schatz or Lehmann. The naming of Waksman as the solitary recipient of the prize must have been a bitter disappointment for Schatz.

After the prize was awarded to Waksman, the Mayo Clinic doctors learned that a senior official of the clinic had told the Nobel visitors that he was not convinced of the validity of the work of Feldman and Hinshaw, and that they had continued their work on tuberculosis against the advice of some Mayo Clinic "higher ups." We cannot be sure of this, but if that were the deciding factor, it would have been a tragic misjudgment.

In a paper published in 1965, Schatz was still very resentful. Nowhere in its eleven pages did he mention Waksman by name, but he made numerous references to his paltry $40 monthly stipend as a graduate student and to his hardships as a student "employee" of an unnamed professor. Schatz believed that he had a strong case in being recognized by the Nobel Committee, but a large segment of the American scientific community, appalled at his lawsuit, continued to support Waksman and almost certainly had a major influence on the committee.

Lehmann too resented the prize being awarded to Waksman. In an article published in 1964, he subtly suggested that it was he rather than Waksman who should be awarded the honor contained in Waksman's Nobel citation as the "discoverer of the first antibiotic effective against tuberculosis." His claim was based on when his animal and clinical trials were initiated rather than when the trial results were published.

Waksman was an extraordinary scholar, researcher, and bibliophile. He was the author or coauthor of more than four hundred scientific papers and numerous monographs. In 1932, he completed a massive compendium entitled *Principles of Soil Microbiology*. The manuscript was refused by several publishers in the belief that that the market for such a tome was too small to make the book commercially viable. To their surprise, the book sold very well, and it dominated the field for two decades. He died at the age of eighty-six, leaving a legacy that is the envy of many scientists. He is buried in the old churchyard at Woods Hole, Massachusetts, beside many of his scientific peers.

Comment

The completely accidental discovery of streptomycin came from a culture plate that was ready to be discarded by Schatz's fellow student, Doris Jones. Schatz, faced with a long list of negative searches from sources in soil, asked his fellow students to give him their cultures from other sources rather than discarding them. This would broaden his search since the other graduate students were studying a variety of organisms from sources other than the soil. Jones gave Schatz a culture taken from the throat of diseased chicken that gave him the mold. This was the culture that he used in discovering *Streptomyces griseus*, which secreted streptomycin into the culture medium that killed a variety of pathogenic bacteria, including the tubercle bacillus. This substance became known as streptomycin. The great step from the laboratory to the bedside of patients with tuberculosis occurred thanks to the work of Feldman and Hinshaw.

Since Selman Waksman was dedicated to lifelong study of soil microbiology, he may have restricted his boundaries of scientific thought and interest to that end. He had two significant opportunities to break out of his narrow vision and break new ground by exploring the clues that were presented to him. In 1932, he received a government grant to study the tubercle bacillus in soil, but he found a more interesting project. Then in 1935, he was shown a culture plate that was a close replica of Fleming's famous plate that showed tubercle bacilli being destroyed by a mold. In the famous Schatz, Bugie, and Waksman paper, none of the authors grasped the implication that the tuberculosis organism was killed by mold. Yet in his Noble Prize address, Waksman quoted from Ecclesiastes 38:4: "The Lord hath created medicines out of the earth; and he that is wise will not abhor them." But Waksman was certainly unwise in ignoring these earlier stunning examples.

Although Waksman received high praise from many quarters, the two heroes of the streptomycin story were the two Mayo Clinic physicians, who conducted conclusive animal studies and successful clinical trials that brought streptomycin to the forefront of the battle against the tubercle bacillus. But how did they get involved with streptomycin in the first place? Feldman's visit to Rutgers in late 1943, initiated by a rumor that Waksman's laboratory was searching for antibiotics, seems to have been the first step in a scientific collaboration that led to spectacular results.

In passing, Corwin Hinshaw and William Feldman should have been strong contenders for the 1952 Nobel Prize, but lost out to Waksman.

In this story, chance played a central role in the discovery of streptomycin. Searching for some new positive results in his vast systematic search of soil molds, Schatz luckily found what he was looking for in a culture taken from a chicken that a fellow student gave to him instead of discarding it. It was pure chance that brought Feldman and Hinshaw together by their sharing a ride from St. Paul to the Mayo Clinic, which led to their intellectual partnership that would play the dominant role in the clinical application of the drug to patients suffering from tuberculosis. During the ninety-mile trip, these total strangers sealed a complementary relationship that has aptly been described as symbiotic. Each had professional skills and experience that exactly fitted that of the other like a hand in a glove. Together, they led the revolution that brought the benefit of streptomycin therapy to millions of sufferers of the dreaded disease of tuberculosis.

REFERENCES

Hinshaw, H. C., and W. H. Feldman. "Streptomycin: A Summary of Clinical and Experimental Observations." *J. Pediat.* 28 (1946), 269–274.

Hotchkiss, R. D. "Selman Waksman. Biographical Memoirs." National Academy Press: Washington, DC, 2003.

Ryan, F. *Tuberculosis: The Greatest Story Never Told.* Eugene, OR: FPR Books, 1994.

Waksman, S., and A. Schatz. "Streptomycin—origin, nature, and properties." *J Am. Pharm Assoc.* 34 (1945), 273–291.

Cephalosporins: Something in the Sewage

The eye sees only what the mind is prepared to comprehend.
　　　　　　　　　　　　—Henri-Louis Bergson, philosopher

Giuseppe Brotzu

Public domain

Giuseppe Brotzu, director of the Instituto d'Igiene de Cagliari in Sardinia, was looking out his window on the top floor of the institute admiring the sandy beaches of Su Succi, a popular hangout for the young people of Cagliari. Cagliari was recovering from World War II, and the city, indeed all of Sardinia, was trying to regain its prewar lifestyle. As he looked down on the beach, Brotzu realized that there was a large effluent pipe that was draining raw sewage into the ocean at a point very close to the swimmers. Brotzu was an expert in the epidemiology of typhoid fever, brucellosis, and tuberculosis, the most common infectious diseases that plagued Sardinia, and he had published many articles on these subjects in Italian public health journals.

Brotzu knew of the thrilling discoveries and successful uses of penicillin and streptomycin in treating serious infectious diseases that had been made during and after the war. As he looked at the beach, he wondered why those young people never contracted typhoid fever, since the sewage contained many pathogenic bacteria, including the typhoid bacillus. Could it be that some undiscovered antibiotic was providing a kind of self-purification of the water near the Su Succi beach? It was a great idea, but it had to be tested. Brotzu, a man of action, went down to the sewage effluent and collected a bucket full of the drainage.

The Sewage Mold

Back in the laboratory, Brotzu spread a small amount on agar plates. After incubating the plates, he saw that an interesting mold had grown that he identified. Brotzu then replanted the mold on a plate at six equally spaced points in a wide circle. At the center, he planted streaks of typhoid bacilli that radiated from the center outward toward the six mold colonies. After incubation, he found that only one streak had grown out to the edge, and that was at a point where no mold had been planted. The mold had killed the other five streaks of the typhoid bacillus. Brotzu's hunch had paid off. He isolated a strain of fungus called *Cephalosporium acremonium* that was producing a substance that inhibited the growth of the typhoid bacillus. Through repeated subcultures, he obtained a concentrated mold juice that possessed powerful antibacterial properties, which he named cephalosporin.

Believing that toxicity trials in animals were unnecessary, Brotzu went straight to patients. He first injected his mold juice directly into staphylococcal and streptococcal boils and abscesses, and they regressed. He next injected the mold juice into several patients with typhoid fever. The patients almost immediately developed severe allergic reactions and high fevers. He reasoned that these unhappy effects were due to impurities in the mold juice. But more importantly, after the acute reactions subsided, the patients showed improvement. Since Brotzu did not have the facilities or the equipment to pursue any further study, he approached the Italian drug companies, but they were reeling from the war and were understandably unwilling to take on a new project of uncertain success.

Frustrated, Brotzu decided to publish his results in a journal entitled *Lavori dell' Instituto d'Igiene di Cagliari* (The Works of the Institute of Hygiene of Cagliari). The journal containing the four-page paper entitled "*La Ricerche su di Nuovo Antibtiotico*" (Research on a New Antibiotic) had almost no circulation, since it was unknown and the paper was in Italian. In the closing sentence of the paper, Brotzu made a plea that "other, better equipped institutes may be able to make greater progress in the selection of the fungus and in the extraction of the antibiotic."

Oxford

In the hope that someone in Britain might continue the work, Brotzu mailed a copy of his paper to an English friend, Blyth Brooke, a British public health officer whom he had known during the war. Brooke was excited enough by the article that he promptly forwarded it to Sir Edward Mellanby, the secretary of the Medical Research Council in London. Mellanby was a well-versed scientist, having done important early work on cod-liver oil and rickets. As secretary of the council, he was a powerful figure in British medical science at the war's end.

Mellanby suggested that Brotzu send a culture of his mold along with his published description of the work to Sir Howard Florey, the man who led the Oxford team through the development and first clinical uses of penicillin. Florey's interest was piqued by Brotzu's account of his discovery. He thought it worth exploring when adequate personnel and facilities were available. In the meantime, the culture was deposited in the repository of cultures that were subcultured periodically in order to keep them viable for possible future use.

In August 1948, three years after the mold cultures had been lifted from the Cagliari sewer drainage, they were quietly growing unnoticed in the same laboratories in which penicillin had been developed. The mold had been recultured many times, since it was the practice of the laboratory. Further work on Brotzu's mold was temporarily eclipsed by the work of Dr. T. L. Su, who under Florey's direction had cultured an organism from the Oxford sewers that produced an antibacterial agent called micrococcin. Micrococcin had attracted Florey's particular attention because of its anti-tuberculosis properties in laboratory tests, but it proved to be too toxic in clinical trials.

Florey was becoming increasingly concerned with the emerging problem of bacteria that were resistant to the available antibiotics. He had a feeling that Brotzu's discovery might be an answer to this problem, and he thought it ought to be developed with this in mind. He asked Edward P. Abraham and Guy G. F. Newton, two colleagues who had been members of the penicillin team, to explore the antibiotic possibilities of the cephalosporin mold juice. It was a great hunch, and it would pay off magnificently. Abraham was the ideal choice for the work; he was a brilliant chemist and Newton was his right-hand man. Abraham was a soft-spoken man, who had an unusual degree of persistence in pursuing any work with which he was involved. He would need it as he undertook this new field of research.

Edward Abraham and Guy Newton

Brotzu was an unknown quantity to the Oxford Group. Abraham conducted a library search to learn more about Brotzu's background. He found many of Brotzu's published papers in Italian medical and public health journals throughout the 1920s and 1930s that dealt with typhoid fever, malaria, tuberculosis, and other epidemic infectious diseases. But a further extensive library search failed to identify the journal in which Brotzu had published his results.

On a visit to Oxford, Brotzu later explained that no other articles were scheduled to appear in later issues, unless he once again turned up something new and interesting. But later Abraham confessed that if the Oxford Group had not received the paper from Brooke, they would have noted with a smile that his paper had appeared in the one and only issue of the journal that had ever been published.

Abraham and Newton worked diligently for six years, culturing and identifying products produced by the mold. In their search, they discovered a new cousin of penicillin, which they called Penicillin N. Abraham wrote to Florey, who was then in Australia, suggesting that since Penicillin N was a great find, any further work on the other substance ought to be abandoned or given over to a commercial interest. He argued that the time and money of the department might be better used for other work. Florey wrote back that the two Oxford scientists should not give up and should try to unravel the mysteries that they had encountered. He advised them to take up the problem as an intellectual challenge. Florey's "sixth sense" was a crucial decision that would yield large dividends.

Abraham and Newton finally isolated the newly found substance and labeled it Cephalosporin C. It had remarkable killing properties against a number of bacteria, including many that were penicillin-resistant. It was a major accomplishment as a weapon against the growing problem of resistant organisms. In 1949 Abraham and Newton filed for patents on their discoveries with an assignment of royalties to the Medical Research Council, even though at the time the long-term commercial future of the drug was in doubt. Florey had been wise in turning over the rights to the council, since in this way the Oxford Group was removed from what could have been mind-boggling legal matters for which they were ill prepared.

Another cephalosporin was isolated that was active against gram-negative bacteria. It was named Cephalosporin N, for negative. One of the cephalosporin families was active against gram-positive bacteria, and was named Cephalosporin P, for positive. (Gram-positive and gram-negative organisms are classified according to the color that they take up on a Gram's stain, blue for positive and red for negative. Although this classification is empirical, it nicely divides organisms according to how they will respond to various antibiotics.)

The American pharmaceutical industry undertook the production, testing, and marketing of the new cephalosporins. The clinical trials and later experiences were very promising, and the drugs were brought to market. A few years later, the sales of cephalosporins soared. The agreement between the council and Abraham and Newton allowed some of the royalties to go to the investigators, but most of it went to supporting research at Oxford. A major beneficiary was Lincoln College, Oxford, of which Abraham was a fellow. It is estimated that Abraham had donated a total £30 million to various activities at Oxford.

The pharmaceutical industry chemists went on to develop four generations of cephalosporins, each generation having several different but closely related drugs. The first generation generally attacked gram-positive staphylococci and streptococci. The second generation's spectrum consists of gram-negative rods, such as *E. coli*. The third generation proved useful against selected members of both gram-positive and gram-negative groups, while the fourth generation has a similar spectrum of activity to the third but is better for attacking resistant organisms. Most of the cephalosporins are active orally. They pass into the fluids of the brain, the eye, and the joints, making them extremely useful for infections in those sites. As in the case of penicillin and streptomycin, organisms resistant to one or other cephalosporins have emerged and will continue to do so.

Comment

Giuseppe Brotzu was already a man of affairs in Cagliari when he made his discovery of cephalosporins. After his discovery, he continued to move up through academic and political circles, holding positions of rector of the university, president of the Sardinian Regional Council, and mayor of Cagliari. Brotzu was a deeply religious man, which he said gave him great concern for the social needs of his homeland and for the many who turned to him for help. In 1971, Oxford awarded Brotzu an honorary degree of Doctoris in Scientia. He was also awarded a certificate from the Medical Research Council entitling him to a share in the royalties of his discovery. Brotzu died in 1976 at the age of eighty-one. Edward Abraham received a knighthood in 1980. He died just two months before the golden anniversary of his definitive work on cephalosporins in 1999. Guy Newton died unexpectedly in the midst of the work on cephalosporins, in 1969.

The role of chance is obvious in this tale. Brotzu's prepared mind snapped at the idea of sewage having a role in preventing typhoid fever.

What he saw was something that others had not seen. His plea for further development of the findings was contained in a short paper that was published in Italian in an obscure journal with a tiny (if any) circulation. Fortunately, his wartime British friend, Blyth Brooke, after receiving the paper, sent it on to Sir Edward Mellanby, who requested that Brotzu send the paper and a culture to the Oxford Group, already famous for their work on penicillin. Florey in turn assigned the project to the very best chemists in his department, Edward Abraham and Guy Newton, who successfully isolated and purified the cephalosporins.

REFERENCES

Bo, G. "Giuseppe Brotzu and the discovery of cephalosporins." *Clin. Micr. Infect.* 6, Suppl. 3 (2000), 6–9.

Hamilton, J. M. T. "Sir Edward Abraham's Contribution to the Development of the Cephalosporin: A Reassessment." *Anti Microb Agents* 15 (2000), 179–184.

Hamilton-Miller, J. M. T. "Cephalosporins: from mold to drug. Sardinia to Oxford and beyond." *J Antimicr Chemother.* 44 (A) (1999), 26.

Lowe, G. "Obituary: Sir Edward Abraham." *The Independent,* May 13, 1999.

CHAPTER 13

Peptic Ulcers: A Discovery in an Empty Lab

The greatest obstacle to knowledge is the illusion of knowledge.
—Daniel Boorstin, historian

Dr. Marshall and Dr. Warren

For some time, Dr. Robin Warren, a staff pathologist at the Royal Perth Hospital, had been observing some mysterious bacteria found near the site of stomach and duodenal ulcers in the biopsies of patients with these disorders. He had found the same organism in patients with chronic inflammation of the stomach lining (gastritis), but they were not found in patients with other diseases of the stomach. Warren was unsure of the significance of his findings, since they violated the orthodoxies of gastroenterology and pathology, which held that no bacteria could survive the acidity of the stomach, and that any bacteria seen in the stomach were the result postmortem change. Dr. Warren's findings could not be taken seriously because of the preceding reasons.

Warren was not the first to see these bacteria, which had unusual shapes that were suggestive of a corkscrew. A search of the literature uncovered numerous other papers beginning in 1889 describing the presence of these bacteria, but none had generated any hypothesis as to their significance. Organized medicine often adopts a hostile attitude whenever its traditional teachings are attacked. This was certainly the case of Warren's discovery. In 1940, Dr. A. Stone Friedberg published a paper showing these bacteria in fresh stomach specimens taken at operation. His professor discouraged any further work, saying, "Maybe, you made a mistake." Had Friedberg been able to pursue this work, it might have led to a much earlier use of antibiotics as treatment for ulcers. Generations of medical students had been taught that an excess of stomach acid generated by stress, cigarette smoking, and dietary factors caused peptic ulcers.

In 1977, Dr. Barry Marshall, a well-trained gastroenterologist, joined the staff of the Royal Perth Hospital. Marshall was looking for a research project when he heard about Dr. Robin Warren's observation of bacteria in the stomach. Marshall was interested, and he visited Warren to learn the details. In their discussions, Marshall heard about Warren's inability to convince others of the validity of his many observations, including some in which the bacteria were found in fresh samples taken by biopsy of living patients.

Marshall was convinced that he and Warren should join in a project designed to study the occurrence of bacteria in patients with various disorders of the stomach. In the work, Marshall would send Warren biopsies from patients with ulcer disease or gastritis that Warren would evaluate as well as samples from other patients having other disorders of the stomach. Marshall would collect clinical information on these patients in order to see if there was any correlation between the findings on the biopsies and the symptoms expressed by the patients.

Over the next several years, Marshall obtained one hundred consecutive biopsies from patients with gastric complaints. The fresh biopsy material was sent to Warren for culture and microscopic examination. Most of the patients had symptoms of peptic ulcer or pain. In collecting clinical information, Marshall asked each biopsied patient to complete a detailed clinical protocol that listed every possible symptom that they might have had. The results were totally unexpected. The presence of bacteria did not correlate with any specific symptom, except perhaps for bad breath and burping. The biopsy reports showed that the corkscrew-shaped bacteria were present in every case of duodenal ulcers and had a significant association with gastric ulcers.

Koch's Postulates

But the fundamental question remained: Do these organisms cause ulcers? Marshall set out to test this question. In doing so, he intended to follow the famous postulates laid down by the great microbiologist Robert Koch in 1890 in his studies of the bacteria that cause tuberculosis. The postulates spell out four requirements that must be met to establish that a given microbe is the cause of a given disease. Koch's postulates are:

1. The microorganism must be found in abundance in all individuals suffering from the disease, but should not be found in healthy animals.
2. The microorganism must be isolated and cultured from a diseased individual.
3. The cultured microorganism should cause disease when introduced into a healthy, disease-free individual.
4. The microorganism must be isolated from the diseased experimental host and must be shown to be identical to the original specific causative agent.

Warren had already shown that the bacteria were present in large numbers in patients with gastritis or ulcer, and of equal importance, they were absent in patients who had other, unrelated diseases of the stomach. Marshall set out to test Koch's second postulate, namely to grow the suspected organism in pure culture. He made many attempts to culture the organisms, but all failed. It was frustrating, since many clusters of the organisms could be easily seen under the microscope, but they were resistant to growth in many different culture media.

Easter Holiday

On April 8, 1982, the Thursday before the start of a five-day Easter holiday vacation, another attempt was made to culture the organism. Like all of the previously failed cultures, this one was treated in the same way that all of the other routine diagnostic cultures taken from other patients in the hospital had been handled, namely that they were incubated for forty-eight hours, removed from the incubator, and evaluated by a technician, whose report was sent back to the doctors and then discarded.

This routine was being followed on this particular weekend. The cultures that were planted on Thursday were to be read by the technicians on the following Saturday. But in their haste to get off on the holiday, they left Marshall's cultures untouched in the incubator. When they returned on the following Tuesday, the cultures were examined. To their great surprise, these cultures showed luxuriant growth! It was easy to show that these organisms were identical to those seen on the biopsy specimens. Koch's second postulate had been satisfied, thanks to the Easter vacation!

Marshall presented these findings at a 1983 Brussels conference that was chaired by Professor Martin B. Skirrow of the Gloucestershire Royal Hospital and a renowned authority. Skirrow was impressed with the work of young Dr. Marshall, who created a sensation when he declared that that people who were free of the corkscrew bacteria would be free of peptic ulcers. It was a brash statement considering that the hypothesis had not been critically tested.

Marshall and Warren sent a paper describing their joint work to the leading medical journal, the *Lancet*, in 1984. Although the editors were willing to publish it, they were unable to find any referees for the peer review, which is the requirement of the journal. Learning of these difficulties in getting the work published, Skirrow first repeated and confirmed the results in his own laboratory and then informed the editors of the *Lancet* of his results. Shortly afterward the paper was published.

In their *Lancet* paper, Marshall and Warren described their success in growing and isolating the unidentified bacillus from biopsies of patients with gastritis or ulcer. These bacteria were able to survive the extreme acidity of the stomach by penetrating the thick mucus lining of the stomach walls where they grew in the grooves between cells. Since the deeper mucus layers are slightly alkaline, the bacteria thrived in a neutral environment protected from gastric acid. The paper also contained a discussion of the inadequacy of acid-suppressing H2-receptor antagonists (the so-called group of H2-RA drugs) that were universally being used to prevent ulcer recurrences. Their paper also contained the hypothesis, which Dr. Marshall would go to great lengths to prove, that these bacteria could be the cause of chronic gastritis and peptic ulcer disease.

The Bold Self-Experiment

The next task was to infect a disease-free host to see if the disease could be reproduced. But here were no known animal models from which to choose. Marshall thought that piglets might be a trial animal model. He inoculated them with a culture of organisms, but none developed the pathological picture of the disease. After seriously pondering the idea of using himself as the experimental host, Marshall decided that he had no other choice. He would be the "guinea pig."

On the morning of the experiment, Marshall skipped breakfast but took 400 milligrams of cimetidine (an H2-RA drug), believing that the organisms would have a better chance of surviving if his level of stomach acid was reduced. Two hours later, he gulped down a cloudy brown liquid that contained a billion *Helicobacter pylori* organisms. For the next three days, he had no symptoms, and he continued his usual activities. But on the third day after a small evening meal, the food felt like a lump of lead in his stomach. On the morning of each of the fifth to the eighth days, he awoke very nauseated and vomited a slimy liquid that seemed to contain no acid. He began to feel very fatigued and to sleep poorly. At that point, his wife told him that he had "a putrid breath." His colleagues had also noticed that during the previous week, but they were too polite to tell him that he had foul-smelling breath.

After ten days, a biopsy of his stomach showed teeming clusters of *Helicobacter*, and at that point, Marshall decided to terminate the experiment. The experiment had succeeded. *Helicobacter* was a proven pathogen! The two biopsies taken on day ten were not enough to define the pathology completely, and he scheduled another biopsy four days later. By then, the vomiting had stopped, and he was symptom-free. The next biopsy, taken on the fourteenth day, showed a picture of healing gastritis. At that point, he began the treatment with an antibiotic, which was continued for a week.

No *Helicobacter* were seen in any of the eight subsequent samples. Cultures, histology, and electron micrographs were all bacteria-free. The *Helicobacter* had been eradicated without any additional treatment. Whatever happened to cause the organism to disappear continues to be a mystery. But Marshall had satisfied the conditions of Koch's third postulate, and since the organisms recovered from his own biopsy during the experiment were identical with those seen in his patients, he had met the conditions of the fourth postulate as well. Marshall and Warren had decisively proven that *Helicobacter pylori* were the cause of gastritis and peptic ulcer disease. In other words, these disorders were really infectious diseases and were treatable with antibiotics.

Natural History of *Helicobacter pylori*

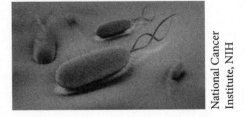

National Cancer Institute, NIH

In pondering the results, Marshall was intrigued by his observations that the vomiting that he had experienced contained no acid. In searching the literature, he discovered some descriptions of gastritis in an account by Sir William Osler in his 1910 *Textbook of Medicine*, in which he describes a similar illness in children in which he had observed a lack of gastric acid. Suddenly the whole natural history of the infection became clear. The reason that ulcer patients could not recall an acute infection with *Helicobacter* was because it had most probably occurred when they were small children, afflicted with some transient illness that was associated with vomiting due to the infection.

Family members might infect these children or others, probably through the oral-intestinal contamination route. Once planted, the *Helicobacter* settled into a lifelong asymptomatic phase, which sometimes was punctuated by the appearance of clinical ulcer disease in adulthood. Because the bacteria were not permanently affected by any of the then customary ulcer therapies, ulcer disease became a lifelong problem, with a pattern of relapses.

Well-controlled double-blind studies from medical centers throughout the world have shown clearly that antibiotic treatment of ulcer patients for a two-week period results in eradication of the *Helicobacter* and healing of the ulcer. Furthermore, the rate of relapse, so common with all previous therapy programs, is nearly zero. One study included heavy cigarette smokers since they were thought to have added risk for ulcers presumably due to stress. But in fact, the level of cigarette smoking remained unchanged in the treated patients. Assessment of the mental status of the treated group showed that their sleep patterns, their sense of optimism, and their feeling of well-being all improved with the eradication of the bacteria. These findings suggest that the so-called "ulcer personality" may simply reflect a diminished state of health related to chronic infection of the stomach.

In 1984, Dutch scientists discovered that *Helicobacter* produce large amounts of the enzyme urease. This enzyme catalyzes the breakdown of urea into ammonia and bicarbonate. The latter probably protects the organism against the acid of the stomach, but in doing so it forms carbon dioxide, which is exhaled through the lungs:

Urea + Urease ==> Ammonia & Bicarbonate
Bicarbonate ==> Carbon Dioxide and Water

Marshall saw the urease reaction as a means for developing a simple, rapid diagnostic test for *Helicobacter*. This test consists of having the patient swallow a capsule containing radioactive-labeled C14 urea and ten minutes later blow up a two-liter balloon to catch the C14 carbon dioxide. The presence of C14, labeled carbon dioxide, in the breath indicates that *Helicobacter* is present. One modification has been to use the non-radioactive carbon isotope C13 labeled urea. The test has proven to be quite effective in confirming the eradication of *Helicobacter* after treatment.

Recognition

Recognition was slow in coming, but it eventually came. Investigators in Australia, Europe, and the United States began to report confirmatory studies. In 1994, the National Institutes of Health convened a consensus panel of experts on "*Helicobacter pylori* in Peptic Ulcer Disease." The publication of this meeting fully supported Marshall and Warren's theory, although it noted that most people infected with *H. pylori* do not develop duodenal or gastric ulcers, implying that some other factors are at work. The consensus report recommended that patients with *H. pylori* ulcers be treated with antibiotics to eradicate the infection, as well as using acid-suppressing agents to relieve the symptoms during the treatment periods.

In 1995, Barry Marshall received the Albert Lasker Clinical Medical Research Award "for his visionary discovery that *Helicobacter pylori* cause peptic ulcer disease." It went on to say "flying against the winds of orthodoxy, Dr. Marshall put the infection theory of ulcers on the map of medical discoveries. The gratitude of untold millions of ulcer patients is embodied in this Lasker Award." The Lasker Award is one of the most prestigious awards for scientists conducting medical research. In many instances, it has been a reliable stepping-stone to the Nobel Prize.

That was ten years in coming, but it came in 2005, when Barry Marshall and Robin Warren jointly received the Nobel Prize in Physiology or Medicine. The citation recognized that they challenged the prevailing dogmas with tenacity and prepared minds. By using technologies generally available (fiber endoscopy, silver staining of histological sections, and culture techniques), they made an irrefutable case that the bacterium *Helicobacter pylori* caused peptic ulcer disease.

Comment

The results of the studies of Marshall and Warren revolutionized the diagnosis and treatment of peptic ulcer disease. Nearly all of the basic findings underpinning their success took place during the early 1980s, yet it took more than a decade before they were fully recognized and appreciated. Dr. Warren's repeated observations of *Helicobacter* were made possible because of the invention of the flexible gastroscope that allowed gastroenterologists to easily take snippets of the stomach lining for microscopic study without undue discomfort of the patient. This was the first time that he could study "fresh" immediate samples from the lining of the stomach or duodenum taken from patients. Despite receiving no recognition of the significance of his work, Warren had persisted.

It was Marshall's curiosity that brought the two together for a long-lasting partnership that had as its foundation the idea of bacteria causing peptic ulcer, which was at the time universally regarded as outrageous. Everyone knew that ulcer disease was caused by a specific "ulcer" lifestyle, the important components of which were stress, cigarette smoking, alcohol, drugs, and perhaps even some genetic factors. Furthermore, orthodox medical authorities "knew" that no bacteria could possibly live in the acid environment of the stomach.

Thanks to Marshall's risky self-experiment, the role of *Helicobacter* in producing ulcer disease was firmly established. Well-controlled studies conducted by Marshall with the essential assistance of Warren were convincing in demonstrating the central role of these bacteria in the genesis of ulcer disease. Looking back, all of the work seems very straightforward. But basically it relied on several unpredictable events. There was the chance visit of Marshall to Warren's laboratory in quest of a research project. It was another stroke of good fortune that Professor Skirrow took up work to confirm the results of Marshall and Warren and then to persuade the editors of the *Lancet* to publish their paper. Then there was the happy circumstance of a long Easter holiday that allowed the *Helicobacter* cultures to grow undisturbed for five days rather than being discarded after two days.

Given the convincing results of the work, what were the sources of opposition to its acceptance? Marshall mentions an economic motive on the part of the pharmaceutical industry. At the time, companies making so-called H2-RA drugs were enjoying sales of $3 billion per year and reaping handsome profits. If these drugs would vanish and be replaced by antibiotics, it might be a serious blow to their financial well-being.

Then there were the gastroenterologists, some of whom may have been receiving "research grants" from grateful drug companies. Furthermore, a significant source of their income was derived from fees for endoscopy procedures and from repeated office visits for adjustments of therapy and for the management of relapses. These income streams for their practices would be markedly reduced, given the simple urease test and the nearly zero rate of relapses. Finally, the gastroenterologist faced potential competition for ulcer patients from general practitioners, who were now armed with antibiotics and the C14 or C13 urease test and could now competently manage their patients with ulcer disease without endoscopy.

Armed with new diagnostic tests for *Helicobacter*, epidemiologists looking at populations on all continents began to see that the bacteria were so widely spread that that it may be the world's most common chronic infection. But even more fascinating were the reports that a decrease in the prevalence of *H. pylori* was strongly associated with a declining incidence of gastric cancer, which in many parts of the world ranks as number two in terms of cancer deaths.

REFERENCES

Cohen, S. "Helicobacter Pylori." *The Merck Manual of Diagnosis and Therapy*, 18th ed. Hoboken: John Wiley & Sons, 2006.

Marshal, B. J. "Helicobacter Connections." Nobel Lecture, Nobel Prize in Physiology or Medicine, 2005, 250–277.

Marshall, B. J. et al. *Helicobacter Pylori in Peptic Ulceration and Gastritis.* Cambridge, MA: Blackwell Science, 1991.

Warren, J. R. "Helicobacter—The Ease and Difficulty of a New Discovery." Nobel Lecture, Nobel Prize in Physiology or Medicine, 2005, 292–305.

Part V
Drugs that Affect the Heart

14

Nitroglycerin and Amyl Nitrite: Boom!

There is a disorder of the breast marked with strong and peculiar
symptoms considerable for the kind of danger belonging to it. . . .
—*William Heberden, 1772*

Introduction

Angina pectoris is a spasm of vise-like pain that occurs whenever the oxygen demand of the heart exceeds its oxygen supply due to an acute insufficient delivery of oxygen-laden blood by the coronary arteries. The underlying cause is arteriosclerosis of the coronary vessels. Angina pain appears suddenly; it is felt over the left chest above the heart, shoots up the neck, and goes down the left arm, often accompanied by shortness of breath and an impending sense of doom. The sufferer is obliged to stop whatever activity in which he is engaged and rest until the pain subsides.

In 1809, David Dundas described a patient seized with angina pectoris as follows: "He was seized with a considerable pain at the heart, and a difficulty of respiration, great palpitation, and great anxiety. He conceived that the smallest motion of the body has instantly destroyed him, and this dread seemed to have totally bereft him of the power of utterance."

Two drugs, nitroglycerin and amyl nitrite, play prominent roles in the development of effective treatment for this disease. In 1844, Antoine Jerome Balard, the discoverer of bromine, was investigating the cause of spoilage of eau de vie de marc, a brandy made from distilled pressed grape pulp that is left over after wine has been made. He identified a volatile oily substance that proved to be amyl nitrite. The vapors provoked a severe headache, but he did not pursue studies of its physiological effects.

English chemist Frederick Guthrie, in studying the compound, wrote: "One of the most prominent of its properties is the singular effect of its vapor, when inhaled, upon the action of the heart. [If a few drops are inhaled after a lapse of fifty seconds, a sudden throbbing of the arteries is felt, immediately followed by flushing and an acceleration of the action of the heart.] These symptoms last for about a minute and then cease." Guthrie attributed these symptoms to a dangerous contaminant, hydrocyanic acid. For this reason, amyl nitrite was regarded as a poison, and no further meaningful studies of its effects were forthcoming until 1865, when Benjamin Richardson (later Sir Benjamin) found that amyl nitrite produced dilatation of the capillaries in the web of a frog's foot, which he saw under a microscope. As an unforeseen sidelight, Richardson also found that the substance had the ability to preserve flowers and dead animal tissue.

Thomas Brunton

A young physician, Thomas Brunton (later Sir Thomas) in Edinburgh, noted that patients with angina pectoris had elevated blood pressure, and he recalled witnessing some experiments of Dr. Arthur Gamgee in which amyl nitrite reduced the blood pressure in animals and humans. Brunton reasoned that both the fall in blood pressure as well as the flushing produced by amyl nitrite were due to dilation of blood vessels. If so, maybe the coronary vessels supplying the heart might be similarly affected, and amyl nitrite should therefore be useful for treating angina pectoris. Brunton had observed a patient in the hospital who was suffering from very severe attacks of angina pectoris lasting an hour every night between the hours of 2 a.m. and 4 a.m. The traditional remedies, such as bloodletting, brandy, digitalis, and even chloroform, brought only transient if any relief, the pain "completely disappeared and did not return till its wonted time next night."

This marvelous result paved the way for the relief of one of the most distressing and excruciating pains that can befall a patient. Amyl nitrite was in common use throughout the latter part of the nineteenth century and well into the twentieth. A common form of dispensing the drug was called the "amyl nitrite pearl." The drug was placed in a small vial that was covered by a protective fibrous white mesh. During an angina attack, the patient broke the vial and inhaled the vapor of amyl nitrite. Nitroglycerin (or one of its derivatives) has now surpassed amyl nitrite.

Nitroglycerin

Nitroglycerin has a different story. In 1850, Alfred Nobel was working in Paris in the private laboratory of Professor T. J. Pelouze, a famous chemist. While there he met a young Italian chemist, Asconio Sobrero, who had qualified as a physician and surgeon but had given up medicine to become a chemist. Three years earlier, Sobrero had produced nitroglycerin by adding glycerin to a mixture of nitric and sulfuric acids. The result was a violent explosion that injured him and several bystanders. He called the substance pyroglycerine (nitroglycerin). Sobrero recalled his reaction to the substance: "Its taste is sweet, pungent, and aromatic. One must always be careful in trying this because a very small quantity (such as in lightly wetting the end of the little finger) on the tongue can bring on a powerful migraine lasting for several hours."

An impressive explosion resulted even when a single drop of the oily liquid was placed on an anvil and struck with a hammer. Sobrero was so shaken by the explosive nature of his discovery that he vowed never to pursue or exploit it. But Alfred Nobel had a great interest in nitroglycerin, believing that it could be put to practical use in construction work, providing, of course, that the safety problems could be solved. Nobel patented nitroglycerin under the name of "blasting oil" and built a manufacturing facility called Nitroglycerin Ltd. in Stockholm. His successes are history.

The recognition of the usefulness of nitroglycerin in medical practice began with the homeopathic movement in both England and the United States. Thanks to a chance conversation that occurred in 1859 between a homeopath and English surgeon A. G. Field, the medical profession was alerted to the unusual properties of small amounts of the drug. Field had used a few drops of a 1 percent solution of nitroglycerin in alcohol in some experiments on himself, and his reactions were summarized in the third person:

> About three minutes after the dose had been placed on his tongue, he noticed a sensation of fullness in both sides of the neck, succeeded by nausea. For a moment or two, there was a loud rushing noise in his ears, like steam passing out of a teakettle. He experienced a feeling of constriction around the lower part of the neck, his forehead was wet with perspiration and he yawned frequently. These sensations were succeeded by a slight headache and a dull heavy pain in the stomach. He felt languid and disinclined for exertion, either mental or physical.

This condition lasted for half an hour, with the exception of the headache, which continued till the next morning.

The symptoms that Field described resulted from a single dose of one-fiftieth of a grain (1.3 milligrams). Thinking that he might be unusually susceptible to the action of the drug, he persuaded a friend to take a dose. This gentleman experienced such decided effects from merely touching his tongue with the cork of the bottle containing the nitroglycerin solution that he refused to have anything more to do with it. Field placed half a drop of the solution on the tongue of a lady suffering from a toothache, and she experienced the same symptoms that he had had. The toothache subsided but she became partly insensible, disliking to be aroused. When she recovered her senses, she complained of a headache, but the toothache was gone.

Others confirmed Field's experiences, but Dr. George Harley challenged Field's results based on his own trials. The controversy revolved around the difference in the strength of the solution that each was using. But when Harley tasted a drop of a stronger solution, he experienced such mild effects that he was inclined to think that the effects experienced by others "were partially due to their imagination." However, when he took a larger dose, he became a partial convert, but he was still more or less convinced that imagination played a large role in amplifying the rather mild symptoms.

William Murrell

Nothing more was made of these early trials until two decades later. Dr. William Murrell, puzzled by the lingering controversy, undertook a self-experiment. He tasted the cork from a bottle of 1 percent nitroglycerin while attending a patient in his consulting rooms. After a few minutes, he was unable to proceed due to violent pulsations in his head, a rapid heartbeat, and fullness throughout his body. He asked the patient to step behind a screen and disrobe in order to gain some time to recover. After several minutes, the worst had passed, but when he bent over to listen to the patient's heart, he again experienced violent pulsations. The symptoms subsided over the next few minutes except for a lingering severe headache. After trying the drug on himself some forty different times, he administered tiny doses to thirty-five of his friends, who reported that their symptoms were for the most part the same as those he had experienced

Murrell surmised that since the action of nitroglycerin closely mimicked the effects of amyl nitrite, it would be useful in treating angina pectoris. The difference was that amyl nitrite was effective within seconds of inhalation, but its effects were short-lived, while nitroglycerin took six or seven minutes to exert its effect, but the effect lasted for up to half an hour. The headache so typical of the earlier experiments was due to too large a dose.

Murrell developed a treatment plan in which the patients would take a few drops in half an ounce of water three times a day to prevent attacks and a similar amount to treat an impending attack. He recommended that "the patient adopt a plan of carrying his medicine with him in a phial and taking a dose if an attack seized him in the street. It never failed to afford relief." The twenty-six-year-old Murrell published his results in 1879 in a classic paper entitled "Nitroglycerin as a Remedy for Angina Pectoris."

Here is his description of one of the first patients he treated with nitroglycerin: "William A. age sixty-four, was complaining of intense chest pain excited by the slightest exertion. He remained well for seven months when another severe paroxysm of pain struck. I ordered drop doses of 1 percent nitroglycerin with great improvement. He now carries a phial of nitroglycerin in his pocket to take if an attack seizes him on the street. It never fails to afford relief."

Comment

There seems to be no direct historical relationship between nitroglycerin as an explosive and nitroglycerin as a drug other than the chance observation of Sobrero in tasting his finger that had been wetted with the oily material. Workers in the Nobel plant that produced nitroglycerin often complained of severe headaches, and some thought that there was an increased incidence of acute heart attacks. Indeed, Alfred Nobel himself had worked in a laboratory with nitroglycerin over a thirty-three-year period, and over the last seven years of his life, he developed signs and symptoms that to some authors were suggestive of chronic nitroglycerin exposure.

These years were punctuated by several heart attacks, frequent bouts of angina pectoris, and intermittent episodes of deep depression. Shortly before his death at the age of sixty-three, Nobel wrote the following letter to his personal assistant: "My heart trouble will keep me here in Paris for another few days at least, until my doctors are in complete agreement about my immediate treatment. Isn't the irony of fate that I have been prescribed N/Gel [i.e., nitroglycerin] to be taken internally? They call it Trinitron so as to not scare the chemist and the public." Nitroglycerin and its derivatives are in wide use today and in various forms. For individual angina episodes, a tablet placed under the tongue or a longer chewable form of a derivative is used. Oral forms of derivatives provide stepwise increasing durations of

effect. Since the skin can absorb the drug in the blood, a nitroglycerin oint-
ment is used for the prevention of angina during sleep. The nitroglycerin
patch provides an even longer-acting preventive that is usually used on a
schedule of twelve hours on and twelve hours off.

REFERENCES

"Heberden's Commentaries on the History and Care of Diseases" published post-
humously in London in 1802 in Murrell W: "Nitroglycerin as a Remedy for
Angina Pectoris." InWillius, FA & Keys, TE: *Classics in Cardiology* 2 (1961),
640–650.

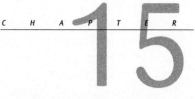

Digitalis: The Shropshire Woman and the Foxglove

Dropsy: A watery fluid collection in tissues of the body.
—Oxford English Dictionary

Introduction

Dropsy is an old-fashioned word for edema, which occurs when fluid accumulates in the lungs, abdomen, legs, and feet. The causes are many, but the most frequent is congestive heart failure, in which patients may accumulate large amounts of fluid. This makes its victim miserable, as is evident from the twelfth-century description by Alexius I Comnenus (1048–1118), ruler of the Byzantine Empire: "[H]is heart they said was inflamed and was attracting all the superfluous matter [fluid] from the rest of the body. . . . Every day it grew worse. . . . He was unable to lie on either side, so weak that every breath involved great effort. . . . He was forced to sit upright to breathe at all . . . but when his stomach was visibly enlarged and his feet also swelled up and fever laid him low, some doctors with scant regard for fever had recourse to cauterization.

"If by chance he did lie on his back or side, the suffocation was awful . . . when sleep in pity overcame him, there was danger of asphyxia . . . the doctors tried phlebotomy and made an incision in the elbow, but that also proved to be fruitless."

Until the dawn of the nineteenth century, dropsy was a scourge (along with consumption), and its effects on its victims were no different than those so described by Alexius I Comnenus five centuries earlier.

Today, the treatment of dropsy due to heart failure relies on one of the modern derivatives of digitalis along with diuretics. This chapter relates how digitalis was discovered some 225 years ago and how it first came into medical practice as a drug for the relief of dropsy.

William Withering: Physician and Botanist

In 1767, William Withering, a recent medical graduate of University of Edinburgh, opened a small practice in Stafford. His early years in practice afforded him the leisure to expand his interests in botany, a subject in which he had distinguished himself at the university. Western England provided a varied landscape and a nearly endless variety of flowers and plants. Aside from pursuing his scientific interest in botany, Withering enjoyed gathering flowers for his future wife for her to paint.

Withering's remarkable knowledge of botany culminated in the publication in 1776 of a two-volume masterpiece entitled *A Botanical Arrangement of All the Vegetables Naturally Growing in Great Britain.* These carefully prepared volumes appeared in English rather than Latin, but they retained the Linnaean system of classification. In discussing the virtues of plants, Withering wrote that the "fables of the ancient herbalists" should be rejected, and he recommended that the subject be approached as altogether new and "built only on the basis of accurate and well considered experiments." Such words reflected Withering's scientific outlook that would become so evident later in his career.

Because his practice in Stafford was generating no more than £100 yearly, Withering moved to Birmingham, although he retained his position as the first physician to the Stafford Infirmary. He had been encouraged to make this move by Dr. Erasmus Darwin, Charles's grandfather, whom he had met. Erasmus Darwin, having learned of the death of a Dr. Small, whose practice had earned £500 yearly, urged Withering to consider moving to Birmingham as his replacement. Withering took this advice and moved to Birmingham in 1775, and he promptly established an active, lucrative practice that grew rapidly. But he also found time to care for many indigent patients. Every two weeks or so, he returned to Stafford, which was some thirty miles away, to look after patients in the infirmary, since Stafford had not been able to attract another doctor as his replacement. It was during one of these trips that Withering discovered digitalis as the key to the effective treatment of dropsy.

The Secret Recipe

Typically, the trip from Birmingham to Stafford, some thirty miles over rough roads, required changing horses midway. At one these stops, a stable worker, who knew that Withering was a physician, asked Withering to examine an older woman, a family member who was suffering from severe dropsy. Withering consented and examined the woman, who was very swollen in the chest, abdomen, legs, and feet from retained fluid. The only treatments available at the time were purging and drainage of the fluid from the swollen abdomen or from the waterlogged feet using small metal tubes. These measures rarely succeeded, and Withering was unwilling to recommend any of these.

One week later, he was returning from Stafford, and he stopped again at the stables for fresh horses. While there, he asked about the woman he had seen previously. Much to his surprise, the woman was now completely free of all the massive swellings of fluid that he had seen before. He asked what had caused this remarkable difference, and the woman replied that she had been given an herbal mixture that had been concocted by an old woman of Shropshire, but the information, other than the existence of the recipe, was a closely guarded secret. Excited by this encounter, Withering launched a detailed inquiry for the contents of the recipe by asking his patients, colleagues, townspeople, and others for any information about this mixture. His efforts were rewarded when he finally obtained a written list of the contents of the mixture.

A painting by Will M. Prince depicts Withering receiving the recipe from an old woman. It has a highly fanciful caption as follows: "Withering exchanging golden sovereigns for the secret recipe held by Old Mother Hutton." Originally, the picture was part of a 1938 advertising campaign directed to doctors by the Parke-Davis Pharmaceutical Company. Little further is known about it, except to point out that two liberties were taken with the real event—notably that Withering did not pay for the recipe according to his records, and the name Old Mother Hutton was almost certainly fabricated. Dr. Dennis Krikler inquired about the history of the picture and concluded that the scene is a myth. Even so, it is a delightful myth.

After studying the recipe, Withering wrote "that this medicine was composed of twenty or more different herbs, but it was not very difficult for one conversant in these subjects, to perceive that the active ingredient could be none other than the Foxglove (Digitalis purpurea)." He gathered some leaves from the plant and used them to brew some tea, which he administered to several of his patients with dropsy with some success. At this point, he was further encouraged by the news that the principal of Brasenose

College, Oxford, had been cured of the "hydrops pectoris," another form of dropsy, by the root of the foxglove. Withering thought the amount of the root that the principal had taken was very large, but being a biennial plant, the root probably varied in composition depending on the season.

Withering gathered the leaves of the foxglove only when the flowers were in bloom, because the potency of the leaves was also variable, depending on the time of the year. Drying the leaves increased the potency fivefold over that of fresh leaves. He added the dried leaves to boiling water to create what he called an infusion. He felt that adding a given amount of dried leaves to a given amount of water created a more precise measured dose.

Over the next ten years, Withering meticulously recorded its successes, failures, and side effects in a 207-page monograph entitled *An Account of the Foxglove and Some of its Medical Uses: with Practical Remarks on Dropsy, and Other Diseases*. The book bore an inscription from Horace, *Nonumque prematur in annum* (and let it be suppressed for nine years). It is ironic that Withering had rejected the "fable of ancient herbalists" in his earlier work on *A Botanical Arrangement of All the Vegetables*, yet he owed his discovery to an herbalist. He did redeem himself somewhat by following his own advice by conducting accurate and well-considered experiments.

Withering's modesty is evident in the preface: "After having been frequently urged to write upon this subject, and as often declined to do so from apprehension of my own inability, I am at length compelled to take up my pen, however, unqualified I may still feel myself for the task." Withering went on to write, "Time will fix the real value upon this discovery, and determine whether I have imposed upon myself and others or contributed to the benefit of science and mankind." Withering did not hold back any ill results in his study of the 105 patients described.

He diligently reported the details of every case regardless of the outcome, good or bad. He described a typical example of the successful use of foxglove in a middle-aged woman with massive swelling: "She was in a state of severe suffocation, her pulse extremely weak and irregular, and her breath very short and laborious. Her stomach, legs and thighs were greatly swollen; her urine was very small in quantity. Five draughts of Foxglove tea acted very powerfully upon her kidneys, and she made eight quarts of urine in 24 hours. Her sense of fullness was greatly diminished, her breathing eased, her pulse became more regular, and the swelling of her legs subsided."

Opposition Arises

Many physicians, some quite eminent, were reluctant to adopt this new treatment for dropsy, probably because they were using too large a dose. The side effects of digitalis overdose were well described by Withering and remain the same today. But the difference between an effective dose and a toxic dose proved to be narrow, which emphasized the need for careful consideration of the doses prescribed. Withering's idea was that digitalis acted directly on dropsy through its diuretic effects on the kidneys. He was led to this conclusion by the fact that if a patient had no fluid retention, he would not respond to digitalis with an increased urine volume. He did recognize "that the drug has the power over the motion of the heart to a degree yet unobserved in any other medicine, and that this power may be converted to salutary ends." Only much later was the action of digitalis in mobilizing the fluid accumulations of dropsy shown to result from a direct stimulating effect on the heart.

One of the early consequences of the emphasis on diuretic action was the use of foxglove in cases of fluid retention caused by liver or kidney failure. In 1806, the famous English statesman Charles Fox developed abdominal dropsy. He was under the care of Dr. John Coakley Lettsom, a highly respected physician with a large and profitable London practice. Withering had written to Lettsom suggesting that he try digitalis on some of his patients with edema. He did so, including Fox, who actually had cirrhosis of the liver with a secondary massive accumulation of fluid in the abdomen. Fox died soon thereafter, and Lettsom thought his demise was hastened by foxglove and concluded that his death was due solely to the new medicine.

A public hue and cry led by Lettsom went up condemning foxglove. In fact, Fox had been a heavy drinker, and his dropsy was due to his liver disease. Others condemned digitalis for its failure to treat water on the brain or fluid in ovarian cysts. But as greater care was exercised in standardizing dosage and in the selection of patients, the value of foxglove emerged as the most valuable agent for managing congestive heart failure. In 1818, Miss Sarah Hoare, whose father had been successfully treated with digitalis for heart failure, wrote the following poem:

The Foxglove's leaves, with caution given,
Another proof of favoring Heav'n,
Will happily display;
The rapid pulse it can abate;
The hectic pulse it can moderate;
And blest by Him whose will is fate,
May give a lengthened day.

Later Developments

The isolation of the active principal of the foxglove leaf proved to be a daunting chemical challenge. In the 1820s, the Société de Pharmacie in Paris offered a 500-franc prize for the isolation of the active principal. After five years, no claimants came forward, and the prize money was doubled. Finally in 1841, two French pharmacists collected the award by isolating a compound that they named digitoxin. It had a highly unusual structure containing a sugar side chain, making it a chemical glycoside. Other digitoxin relatives were also found, and the group became known as cardiac glycosides.

In the 1920s, the powdered leaves of *Digitalis lanata*, the "woolly foxglove," were found to have greater activity than those of *Digitalis purpurea*. An active principal different from the others was isolated and named digoxin. Digoxin is now the cardiac glycoside of choice, since the required blood levels are achieved more rapidly and the drug is more quickly excreted from the body. In 1888, Léon-Albert Arnaud isolated yet another cardiac glycoside from the bark and roots of the ouabaio tree that the Somalis of East Africa used for preparing their arrow poison. The active principal was easily isolated and named ouabain. Because of its rapid onset of action, it proved to be very useful in selected clinical situations. Finally, to add even more esoterica, another cardiac glycoside was isolated from squill, the bulb of the sea onion. The ancient Greeks and Egyptians knew this as the source of an expectorant.

Withering's *An Account of the Foxglove* is a classic that is highly prized among collectors. When first published, it sold for five shillings and included a colored plate of *D. purpurea*. In 1943, a copy fetched $270; today, it would command a far greater price. His two-volume masterpiece on *A Botanical Arrangement* went through multiple editions; the fourteenth and final one appeared in 1877, a hundred years after the first version was published.

Comment

Withering's health began to fail at age of forty-four due to tuberculosis. The disease progressed over the following fourteen years and claimed his life at the age of fifty-eight. After his funeral, a friend wrote that the "flower of English botany was Withering." Stopping midway during his trips to Stafford by chance, he learned of a woman whose dropsy had been amazingly cured by following the secret recipe of an old woman of Shropshire. Thanks to his knowledge of botany, Withering immediately spotted the foxglove as the active ingredient among the twenty components in the recipe. Withering was not only an excellent clinician, but he was also a good clinical scientist. He took careful notes of each patient treated with the foxglove, whether the results were good or bad, and he waited nine years before publishing his results.

REFERENCES

Cushny, A. R. *Action and Uses in Medicine of Digitalis.* London: Longmans, Green and Co., 1925.

Krikler, D. "Withering and the foxglove: the making of a myth." *Br Heart J.* 54 (1985), 256–257.

Lutz, J. E. "Alexius I Ruler of the Byzantine Empire: An XII century description of congestive heart failure." *Am J Cardiology.* 61 (1988), 494–495.

Norman, J. M. "William Withering and the Purple Foxglove: A Bicentennial Tribute." *J. Clin. Pharm.* 25 (1985), 7.

16

The Tranquilizers:
Two Chance Discoveries

Cricket is basically playing baseball on Valium.
—*Robin Williams, actor*

Introduction

In 1945, Frank Berger, a pharmacologist, and William Bradley, a chemist, were working at a British pharmaceutical company, attempting to perfect a disinfectant that would be effective against bacteria that were resistant to penicillin. By manipulating the chemical groups on an existing compound, they produced a new compound, but it had only slight antibacterial activity. Before discarding it, they tested it on animals and were surprised to find that it produced a limp paralysis of muscles without loss of consciousness or reflexes. After a short time, the animals recovered completely. The agent seemed to be a good muscle relaxant.

Following encouraging animal and human trials, the new compound, named mephenesin, was introduced into clinical medicine in 1947 as a muscle relaxant for anesthesia. Mephenesin also had anti-anxiety properties without clouding consciousness. Several drawbacks, however, soon became apparent. The drug had a very short half-life in the body; it had undesirable effects on spinal nerves; and large doses were required to achieve the anti-anxiety effects.

Some molecular modifications eliminated these effects, so that by 1950, a new drug called meprobamate was created. The trade name of the drug was Miltown, after Milltown, New Jersey, where it was first produced and tested. (Note that the drug Miltown has only one l, while the town it was named after has two.) Miltown had a duration of action much longer than mephenesin, it was readily absorbed from the intestinal tract, it produced muscle relaxation, and it relieved emotional tension. It was the first of the tranquilizers. One of the most dramatic effects of Miltown was its remarkable effects in calming raging lions and tigers.

Even though the drug was not widely advertised, it set off a wave of widespread popularity so that by 1957, two years after its release, 36 million prescriptions had been filled, amounting to a billion of so-called "peace pills." The drug's popularity plummeted when it was realized that it was causing widespread psychological dependence. Even so, Miltown set the stage for any drug that would be readily accepted if it provided psychological relief in the "age of anxiety," the self-proclaimed era that Americans were told was gripping their collective psyche.

The Discovery of Valium

A group of improved tranquilizers, called the benzodiazepines, were about to be discovered, and they would completely eclipse Miltown. Dr. Leo Sternbach was the man who made it all happen.

In the 1930s, Dr. Sternbach was working as a graduate student in the University of Cracow's chemistry department. His PhD thesis dealt with a class of unusual chemical compounds that had seven membered-ring structures. He learned how to synthesize these compounds, and he studied their theoretical properties, but he had no interest in their possible biological activities.

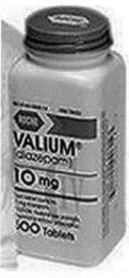

Public domain

In the 1950s, Sternbach, now director of medicinal chemistry at the Hoffmann-La Roche Company in New Jersey, had been trying to produce a tranquilizer to compete with Miltown. He had tried many modifications, but all twenty-three had tested negative in animals. Because of the long run of negative results, a senior manager canceled all further work on the project, although at the very end of the project, Sternbach added another compound, number twenty-four to the list of twenty-three, and he labeled it simply Ro 5-0690. It was put on the shelf along with the others and attracted no further attention.

During a laboratory cleanup some months later, all twenty-four vials in the labeled "inactive" group were slated to be discarded. But before throwing them out, Dr. Sternbach's assistant at the last minute pulled twenty-four off the shelf and told Dr. Sternbach that Ro 5-0690 had never been tested. Dr. Sternbach responded by saying, "Let's try it on the 'cat test.'" The cat test involved picking up a cat after injecting the drug, suspending it by the nape of its neck, and observing whether it was limp or not. To the surprise of everyone, number twenty-four was positive. The cat was limp. More animal tests followed.

But after the cat test was the "Leo test." Dr. Sternbach always tested each drug that he had produced by taking it himself. The "Leo test" was definitely very positive. He was elated that his still no-name compound had given him a feeling of well-being and relaxation without any side effects. (Later Dr. Sternbach abandoned the "Leo test" following a bad reaction to another test.)

All attention in the laboratory was refocused on Ro 5-0690. Dr. Sternbach proceeded to develop the first generation of a family of chemicals that have come to be known as the benzodiazepines, the first of which came to market under the trade name of Librium. Sufficient experience had accumulated by the end of 1959 to warrant holding a symposium on the drug, and a few months later the FDA approved its use. One month thereafter, the drug was marketed under the name Librium. Its immediate acceptance by physicians and patients seemed to meet a large, unfulfilled need for a drug that reduced anxiety.

Research on the benzodiazepines quickened after the success of Librium. Dr. Sternbach synthesized fifty-five different derivatives, and each was tested. One proved to be five times more potent than Librium in animals. This drug was named Valium, from the Latin *valere*, meaning "to be strong." Even though diazepam was the drug's generic name, the name Valium caught on.

Clinical Experiences

Valium was set for evaluation in clinical trials. The very first trial involved elderly patients who were given large doses. Because they experienced slurred speech and unsteady gait almost immediately, the trial was stopped, and the drug was withdrawn from any further study. Dr. L. R. Hines, director of biologic research for Hoffmann-La Roche, realized that the doses were excessive and the elderly were probably the wrong patients for the initial study.

Hines launched new trial, selecting only those patients who most often visited a psychiatrist's office. In a very short time, the drug proved effective in controlling tension and anxiety with a minimum of side effects and without clouding consciousness or interfering with intellectual acuity. This initial experience was considerably expanded to an evaluation of the drug in prisons, clinics, and private offices. All concurred with the earlier findings. Valium was safer and more effective than any of the previous sedatives such as barbiturates, opiates, alcohol, and herbs.

Dr. Irvin F. Cohen, one of the early clinical investigators, pointed out that the history of the discovery of the first benzodiazepines contained all of the elements of high drama. It was marked with "exploration of the unknown, disillusionment and disappointment, abandonment of the project and rediscovery of a discarded compound, excitement in the laboratory followed by discouragement from the clinic, and eventual confirmation of the faith of early supporters."

Valium Sales Spiral Upward

Approved in May 1963, Valium became immensely popular in an extremely short period of time, and was the country's most prescribed drug during the period from 1969 to 1982. Peak sales hit $600 million a year, grossing hefty profits for Hoffmann-La Roche. Valium became a cultural icon; it was a big hit among the anxiety-ridden public. For some people, it was "Executive Excedrin;" for others, it was "Mother's Little Helper," which the housewife used to get her through her busy day. By 1978, Americans were ingesting 2.3 billion of the little white, yellow, or blue pills with a "V" in the shape of a heart stamped on them.

Following Valium's meteoric success, many more different benzodiazepines were synthesized. In some instances, the aim was to identify a certain group on the basic structure that would enhance one of the effects of the drug without eliminating others. Some of these agents are among the twenty-one benzodiazepines that have survived to the present.

As Valium became a part of the family medicine chest, cases of addiction to the drug mounted. The alarm that grew over the many abuses grew to the point where a US Senate committee held hearings that were prompted by celebrities who were addicted to the drug, often taken along with an excessive intake of alcohol. Elizabeth Taylor was hooked on Valium and Jack Daniel's; Tammy Faye Bakker was fond of Valium and a nasal spray cocktail; and Michael Deaver, a highly placed aide to President Ronald Reagan, said he lied to a grand jury because he was in a "Valium haze." Sales of Valium plummeted from number one to number 189 on the list of most-prescribed drugs.

Comment

Dr. Sternbach retired from Hoffmann-La Roche in 1973. In an interview in the late 1990s when he was ninety-three, he remarked that he still couldn't understand all the flak about his most celebrated discovery. While acknowledging the addictions, he pointed to the positive things that Valium had achieved—preventing suicides, healing broken marriages, and the like.

The discovery of Miltown in 1950 and the discovery of Librium and Valium shortly thereafter are great examples of a chance observation and the sagacity of the observer to act on it by taking up a new path of inquiry. Dr. Frank Berger's failed search for a new disinfectant led by chance to some unexpected results in animals. But he had the foresight in viewing these surprising findings to change his course of research course entirely, which led to the discovery of Miltown.

In case of Valium, a laboratory assistant informed Dr. Leo Sternbach, his chief, that he was about to discard a compound that had never been tested, even though it was a part of a group of twenty-three others that had already been shown to be inactive. Sternbach's mind, trained and experienced for years in pharmaceutical chemistry, and knowing the importance of paying meticulous attention to detail, ordered the tests. He was surprised and pleased with the positive test results that ended with the discovery of the benzodiazepine family of drugs, of which Valium became the blockbuster.

REFERENCES

Baenninger, A. *Good Chemistry: The Life and Legacy of Valium Inventor Leo Sternbach*. New York: McGraw-Hill, 2003.

Dokoup, T. "How Mother Found Her Helper." *Newsweek*, January 22, 2009.

Tone, A. *The Age of Anxiety*. New York: Basic Books, 2009.

Wikipedia, the Free Encyclopedia. "Diazepam."

LSD: A Mind-Blowing Accident

"Medicine for the soul"
                    —Albert Hofmann, 1980

Introduction

The discovery of the hallucinogenic properties of LSD from ergot is a frightening chapter in medical history. For centuries, ergot, a parasitic fungus (*Claviceps purpurea*) that contaminates rye in the field, plagued Europeans who ate bread made from contaminated rye flour. The victims developed gangrene, convulsions, miscarriages, and a severe burning of the skin, called St. Anthony's fire. St. Anthony, a religious Egyptian hermit, was the founder of Christian monasticism in the third century. His popularity as a saint reached its height during the middle ages when the Order of Hospitallers of St. Anthony was founded near Grenoble, France, in 1100. It became the shrine for pilgrims seeking relief from ergotism, known then as St. Anthony's (Holy) Fire.

Albert Hofmann's Self-Experiments

Ergot was of great interest to the chemists working at Sandoz Research Laboratories in Basel, Switzerland. They isolated lysergic acid and modified it to produce a drug now called ergonovine, which has powerful effects on uterine contractions. In 1938, Dr. Albert Hofmann combined lysergic acid (LS) with diethyl amide and produced diethyl lysergic acid, adding the "D" to the "LS" and creating the name LSD. After some animal experiments, interest waned, only to be revived in 1943 by Hofmann, who thought it deserved further study. An accidental observation led to an ill-planned self-experiment with a dramatic result. Hofmann later prepared a detailed report, part of which follows:

> Last Friday, April 16, 1943, I was forced to stop my work in the laboratory in the middle of the afternoon and to go home, as I was seized by peculiar restlessness associated with a sensation of mild dizziness. On arriving home, I lay down and sank into a kind of drunkenness which was not unpleasant and which was characterized by extreme activity of imagination.

> As I lay in a dazed condition with my eyes closed and as I experienced daylight as disagreeably bright, there surged upon me an uninterrupted stream of extraordinary plasticity, accompanied by an intense, kaleidoscopic-like play of colors. This condition gradually passed off after about two hours.

Hofmann had been working that afternoon with LSD in a water-soluble form. But he was completely perplexed as to how LSD might have gotten into his body. (Later, a lab assistant commented that a drop of the LSD solution with which he had been working had very likely contaminated a glass rod, which Hofmann used to stir his tea.)

He knew his symptoms were different from those of an ergot overdose, and he devised an experiment on himself to get at the root of the problem. He chose the lowest dose of LSD based on the dosages of other drugs that had been derived from ergot. He was in for a big surprise. His laboratory records on April 19, 1943, note that at 4:20 p.m., he ingested 1/25th of a milligram in water and noted that it was tasteless. At 5:00 p.m., he experienced slight dizziness, unrest, difficulty in concentrating, visual disturbances, and an uncontrollable desire to laugh. At this point he was unable to continue any further writing. He asked his laboratory assistant to accompany him on the bicycle trip to his home.

During the trip, he was unable to speak coherently, his vision swayed, and distorted images appeared like those on a curved mirror. He felt he wasn't moving, although the assistant later told him that he was cycling at a good pace. A physician was called, and excerpts from Hofmann's notes are as follows. By the time the doctor arrived, the peak of the crisis had already passed:

> As far as I can remember, the following were the most outstanding symptoms: vertigo, visual disturbances; the faces of those around me

appeared as grotesque colored masks; marked motorist [sic] unrest alternating with paralysis; an intermittent heavy feeling of the head, limbs and the entire body, as if it were filled with lead.

There was a dry, constricted sensation in the throat; feeling of choking; but a clear recognition of my condition, in which state I sometimes observed in the manner of an independent, neutral observer, that I shouted half insanely, babbling incoherent words. Occasionally, I felt as if I were out of my body.

The doctor found a rather weak pulse but an otherwise normal circulation.

Six hours after the ingestion of the LSD, my condition had already improved considerably. Only the visual disturbances were still pronounced. Everything seemed to sway, and the proportions were distorted like the reflections in the surface of moving water. Moreover all objects appeared in unpleasant, constantly changing colors, the predominate shades being a sickly green and blue. When I closed my eyes, an unending stream of colorful, very realistic and fantastic images surged in upon me.

A remarkable feature was the manner in which all acoustic perceptions, such as the noise of a passing car, were transformed into optical effects, every sound evoking a corresponding colored hallucination, constantly changing in shape and color like pictures in a kaleidoscope. At about one o'clock, I fell asleep and awoke the next morning feeling perfectly well.

This first systematic experiment with LSD was dramatic and important. Volunteer colleagues of Hofmann promptly confirmed the effects but with doses one-fifth of the amount that Hofmann had ingested. LSD was by far the most potent hallucinogenic yet known, being 10,000 times that of mescaline, which produces similar effects.

Mysteries of LSD

To give the reader a sense of how small a dose of LSD is required to produce hallucinations, consider a single grain of sugar under a microscope. It weighs about a milligram. If one divided that single grain into ten parts, yielding 1/10th of a milligram each, one would have the equivalent size of a dose of LSD. Jean Thuillier in his book, *Ten Years that Changed the Face of*

Mental Illness, uses this analogy and writes that such a tiny amount would "in a few moments empty your head of daylight and the notion of space and replace your normal reason with fantasy or anguish, euphoria and a distorted view of the world."

One of the deepest mysteries about LSD is that the brain contains many thousands more nerve cells than the number of LSD molecules being delivered to them. Clearly, LSD must act on only a few specific brain sites, and at these sites, there must be biochemical reactions that underlie this "artificial psychosis" that it produces. LSD provided a whole new way of looking at the causes of psychoses. No longer could they be regarded as purely "psychic" in origin. LSD raised the question of whether tiny amounts of other chemicals in the brain are the basis of normal and abnormal mental states.

Today this is a well-established concept, and a great deal of research is being conducted to understand it. Hofmann's work was exciting, both chemically and medically. Many chemists turned their attention to research on LSD and other hallucinogens. Psychiatrists experimented with LSD on their patients and on themselves. It was a way of fabricating mental illness. Many thought that these experiments provided them with a clearer understanding as to how to approach problems in their patients.

LSD Use Spreads

Aldous Huxley was an important figure in the early history of LSD. Huxley was famous in literary circles and especially noted for his novel *Brave New World.* He described his experiments with psychedelic drugs in his writings, in particular in his book *The Doors of Perception.* His writings helped spread the awareness of psychedelic drugs to the general public.

Timothy Leary

Philip H. Bailey

The tales of LSD hallucinations spread quickly and attracted many nonprofessionals, of whom Timothy Leary became the most known. Leary, who taught psychology at Harvard, called LSD and other hallucinogenic agents "cerebral vitamins." He regarded LSD as a nutrient for the mind; if it were lacking, the world could not be seen properly and true understanding could not be attained. He urged Harvard students to do their own experiments with LSD, and many did. Leary was asked to leave Harvard and eventually was forced to leave the country for Acapulco, where he set up a boarding house for anyone who could buy LSD and administer it to themselves at their whim. He returned to upstate New York and set up a foundation, to which hundreds of young people flocked, meditated, and took LSD. He settled in California, where he learned that he had prostate cancer. On the night of May 31, 1996, the seventy-five-year-old Leary suddenly sat up in bed and shouted, "Why not? Why not? Why not?" He slumped over, dead. He had made sure before that the entire event was videotaped.

In 1975, the public was outraged to learn from the testimony during congressional hearings that revealed that a CIA secret research project involving LSD with the code name of MK-ULTRA had existed for nearly twenty years. Information about many details of the program was lacking because all the files on the project had been destroyed in 1973 on orders of the CIA director. Yet enough evidence was available to reveal an alarming picture. The basic plan was to monitor behavior patterns of unwitting citizens at all levels of society after they had received doses of LSD without their knowledge. Thirty universities and institutions were involved in the extensive testing program, even though many of the monitors of behavior were unqualified. The overall aim of the program was to provide leads for developing methods for effective interrogation and "brain washing" of enemy spies and the like.

Controlling the Spread of LSD

The nation was becoming alarmed at the spread of LSD and especially at the effects it was producing on young people. As the number of drug traffickers and addicts grew, pressures mounted for something to be done. Sandoz, the manufacturer, responded by announcing a new policy under which it would no longer supply any LSD to any scientist or to anyone else. All research on the drug came to an abrupt halt. Sandoz had never marketed the drug but had always supplied it gratis to creditable scientists for research.

Although this ban stemmed the legitimate LSD supply, it did not stop underground laboratories from producing large amounts for resale on the street. Because an active dose of LSD is tiny, a large number of doses can be made from a relatively small amount of ergotamine tartrate as the starting material. The absolute quantities of the drug offered for sale are small, making it easier to hide, carry, or smuggle LSD than any other illegal drug.

The illegality of LSD is still a problem that is with us, despite the provisions of the 1965 Drug Abuse Control Amendments. This statute mandated that the manufacture, sale, transfer, and use of LSD be highly restricted. When Sandoz stopped manufacturing LSD, it turned over its remaining stock to the National Institute of Mental Health, whose policy is not to release any to anyone. The military saw LSD as a possible psychochemical agent in warfare, and it is maintaining a separate stock of the drug, but under very tight controls.

Comment

Many in the psychiatric community have expressed opinions on the very stringent restriction on the use of LSD. It accelerated the widespread illegal and socially undesirable use of the drug. But the inability to use LSD as a research tool for analyzing and understanding the human mind is a distinct negative. Much could be learned about mental disease from LSD studies, and in addition it might be a tool for better understanding of the creative processes of artists, poets, philosophers, and scientists. Jean Thuillier raised the provocative question: "Who could deny that a better understanding of the mind's infrastructure is useful?"

REFERENCES

Hofmann, A. *LSD: My Problem Child*. New York: McGraw-Hill, 1981.

Thuillier, J.: *Ten Years that Changed the Face of Mental Illness*. Cambridge, MA: Blackwell Science, 1999.

Wikipedia, the Free Encyclopedia. "History of LSD."

18

Lithium: The First Guinea Pigs . . . Were Guinea Pigs!

The brain, like any other organ, can get sick and it can also heal.
—Dr. John F. J. Cade

About Lithium

Lithium is a member of the same chemical family as sodium and potassium, and like them it is widely distributed in nature. In 1876, Sir A. B. Garrod introduced lithium in medicine for the treatment of gout. Gout is caused by the deposits of highly insoluble uric acid crystals in joints, especially in the great toe and the kidney.

Garrod found that the uric acid salt of lithium, lithium urate, was the most soluble of all urate compounds, and he reasoned that if cartilage containing urate deposits were immersed in a solution of lithium carbonate, the urate would dissolve more readily than in solutions of sodium or potassium carbonate. The observation that lithium urate is soluble provided a key to Dr. John F. J. Cade's experiments as discussed later.

Lithium salts have been used in a wide variety of ailments, but toxicity was often seen. During the 1940s, lithium chloride was popularized as a substitute for sodium chloride in patients with edema who retained salt because of heart failure or other causes. It was used in an uncontrolled way, and inevitably, disastrous effects followed. In 1949, the *Journal of the*

American Medical Association published reports of deaths due to lithium toxicity, and lithium use fell completely out of favor.

Dr. John Cade

The use of lithium to treat certain psychiatric disorders came about as a result of the work of a psychiatrist, Dr. John Cade. Following his release in 1946 after three years of privation from a Japanese prison camp, the thirty-seven-year-old Cade took up the duties of medical superintendent and psychiatrist at the Repatriation Mental Hospital, a small mental hospital in Bundoora, Australia, a town about thirteen miles from Melbourne. In 1949, Dr. Cade discovered some unrecognized effects of lithium through an unusual set of circumstances.

The key to Cade's discovery came as a result of his clinical curiosity about a possible relationship between manic patients and the heightened (manic-like) activity that is seen in patients with an excessive secretion of thyroid hormone. Manic-depressive insanity was analogous to states of hypo- and hyperthyroidism, with mania being a state of intoxication by excess of some circulating factor, while melancholia is the corresponding state of deprivation. He began his search for the hypothetical toxic agent. He reasoned that in manic states, the body was producing excessive amounts of some intoxicating substance that might be excreted in the urine just as in cases of overacting thyroid, where thyroid hormone is produced in excess. Cade may have lacked adequate facilities, but he did have the freedom to pursue the work in his own way that was unhindered by criticism or caution. Much later he remarked, "These days, one would be suffocated by hospital boards, research committees, ethical committees and heads of departments. Instead, I was answerable only to my own conscience and personal drive."

Experiments in Animals

Having no laboratory facilities, Cade adapted a small deserted pantry for his animal work. He injected fresh concentrated urine from manic patients into the guinea pigs and found that it was more toxic than the urine from non-manic patients, which he used as controls. The latent period after injection of the toxic urine was twelve to twenty-eight minutes, after which the animals became tremulous and uncoordinated, followed by paralysis, convulsions, and death.

In the search for the toxic compound, Cade first thought of urea, the end product of protein metabolism and a major component of urine. Urea injections from manic patients proved to be toxic, with animals being killed by much lower amounts than with the urine from patients with other psychiatric conditions. But a complicating problem arose when he discovered that the amount of urea needed to provoke the animal syndrome very far exceeded the amount of urea in the manic patients' urine. He therefore reasoned that there must be other substances in the urine that modified the effects of urea. Uric acid was a possibility, but given its notorious insolubility, the choice was limited to the lithium salt of uric acid, one of the very few urate salts that are water-soluble. He injected urea in both groups of guinea pigs. In one group he injected lithium urate, while the second group received an equivalent amount of lithium carbonate and served as controls. Both groups became lethargic and unresponsive, but were otherwise fully conscious. It was the lithium that was producing these effects!

Trials on Patients

Cade was eager to try lithium on his patients. Uncertain of the dose, he took lithium and, experiencing no side effects, deemed it to be safe. He selected ten manic patients as the first group to receive lithium. The first of these was "a little wizened man of fifty-one years who had been in a state of chronic manic excitement for five years . . . amiably restless, dirty, destructive, mischievous, and interfering. . . ." The patient "enjoyed preeminent nuisance value in a back ward . . . and bid to remain there for the rest of his life." Lithium treatment began on March 29, 1948. On April 1, Cade thought he saw signs of improvement but conceded that it could have been his "expectant imagination." However, by the next day the patient was clearly "more settled, tidier, less disinhibited, and less distractible." The patient continued to improve and subsequently returned to his old job. The results with the other nine manic patients were "equally gratifying." Within five days, the other manic patients became more settled, tidier, and less distractible, and by three weeks, their behavior was perfectly normal. Some patients with schizophrenia responded very well to lithium, but those with delusions did not respond at all. But in the patients with typical manic-depressive illness, the results were most marked. A maintenance program was developed so that lithium could be safely given repeatedly for patients who were discharged. Looking back, Cade remarked, "It seems a long way from lethargy in guinea pigs to the control of manic excitement in patients."

In 1949, Cade published his first results in a paper entitled "Lithium Salts in the Treatment of Psychotic Excitement" in the *Medical Journal of Australia*. The paper did not arouse an immediate interest in the US and European medical communities. Later, Cade, who was a self-effacing man, speculated that a discovery "made by an unknown psychiatrist with no research training, working in a small hospital with primitive techniques and negligible equipment was not likely to be compellingly persuasive." He might have added that at the same time—1949—the medical public was reeling from the deaths of cardiac patients who had used lithium chloride as a substitute for sodium chloride for patients on a low salt diet.

Mogens Schou

But in the early 1950s, a young Danish psychiatrist named Mogens Schou read Cade's paper and kept the lithium flame alive. Schou continued to champion lithium as a useful agent for four years. Little by little his message caught on, but it still fell on many deaf ears. Schou's great contribution was the results of his carefully designed double-blind placebo-controlled study that fully confirmed Cade's findings. He found that the blood levels of lithium were critical in producing favorable results while avoiding toxicity.

In the 1960s, Poul Christian Baastrup and Schou made sporadic observations that suggested that lithium had prophylactic properties in manic-depressive illness. The idea that lithium could prevent recurrences was an exciting new breakthrough, but it met great resistance by the then orthodox psychiatric community. Their work was labeled "dangerous nonsense" and "a therapeutic myth." Stung by such criticism, Schou and Baastrup undertook a unique double-blind trial with a random allocation of manic-depressive patients to lithium or placebo. It was an unparalleled trial design that had never before been done in psychiatry. The results, which fully confirmed their hypothesis, were published in the *Lancet* in 1970.

For much of the rest of his life, Schou worked diligently and enthusiastically to make the lithium treatments available to all those in need. The spread of lithium use during this period can be estimated by the number of lithium-related published papers in the world medical literature. In 1963, thirteen publications on lithium appeared in the worldwide medical literature; from 1964 to 1968, the number had risen to thirty; but by 1975, the total had risen to more than three thousand.

It took some time before the psychiatric communities in Europe and the United States fully accepted lithium as a new treatment for manic depression. Many psychiatrists simply could not accept the idea that a simple chemical like lithium could radically change the behavior of manic-depressive patients. Neurobiochemistry was unknown at the time. The prevailing psychiatric theory was that psychosis or neurosis had some vague physical basis, but it was certainly not a chemical one. Two distinguished British psychiatrists led the opposition to lithium. One was Sir Aubrey Lewis, professor of psychiatry at the Institute of Psychiatry, whose influence was thought to have been responsible for the blooming of British and European psychiatry after World War II.

The other, Michael Shepherd, professor of epidemiological psychiatry at the Institute of Psychiatry, was regarded as one of the most influential and internationally respected psychiatrists at the time. The opinions of these two experts were based entirely on their mindset that no chemical could possibly play any role in mental illness.

Yet neither of these two gentlemen had ever even tried lithium in their practices. That breakthrough was primarily due to the monumental and persistent efforts of the Danish psychiatrist Mogens Schou.

Comment

Lithium has a double action. It corrects psychological and mental excitation while preventing expansive and depressive derangements. It is preventive against relapses of mania or depression, and it is unique and different from all other psychotropic drugs. Its action has been characterized as a "psychological function regulator," or simply a "mood regulator."

Cade's seminal paper published in the *Medical Journal of Australia* in 1949 sparked a revolution in the thinking of the psychiatric community about the treatment of mental illness. But the revolution took more than twenty years before it was recognized worldwide. Mogens Schou did much to spread the word of lithium's success. His research was of great importance, since it was based on the results of scientifically designed clinical trials.

Cade discovered lithium, a cheap, naturally occurring, and widely available substance, by conducting some elementary experiments in animals. His discovery was another instance of an accidental chance finding by a discoverer who had the sagacity to recognize its importance. Stimulated by these observations, Cade developed a new approach to the treatment of manic-depressive disorders. For countless people around the world, lithium use has meant the ability to regain and maintain mental stability. In 1985, the National Institute of Mental Health estimated that the use of lithium in treating manic depression has saved the world an estimated $17.5 billion. Cade's discovery of lithium made it the first psychotropic drug, and it revolutionized thought on the biochemistry of psychiatric diseases, a field that at the time was completely foreign to the then prevalent ideas about the cause of psychiatric diseases.

It is important to note that although Cade is properly credited with the discovery of lithium, his original three-page paper could have easily been ignored had it not been for impressive work of Mogens Schou and colleagues, who presented irrefutable clinical studies that demonstrated the effectiveness of lithium.

Cade and Schou both received many honors for their work on lithium. They shared the Kittany Award, which at the time was the world's most prized honor in psychiatry. In 1999, the fiftieth anniversary of Cade's discovery of lithium, the Australian government dedicated the John F. J. Cade Institute.

REFERENCES

Johnson, F. N., and S. Johnson. *Lithium in Medical Practice.* Lancaster: MTP Press, 1978.

Mitchell, P. B., and D. Hadzi-Pavlovic. "Lithium treatment for bipolar disorder." *Med. J. Austral.* 171 (1999), 292–264.

Schou, M. "Lithium in psychiatric therapy and prophylaxis." *J. Psych Res.* 1968 6: 67–95.

Schou, M. *Lithium Treatment of Mood Disorders: A Practical Guide.* Basel: S. Karger, 2007.

Thorazine: A Cure from Curare

Schizophrenia is the most distressing, mysterious, surprising, abstruse,
moving, impenetrable, disturbing and obscure of all mental illnesses.
—Dr. Jean Thuillier, 1999

Dr. Henri Laborit

I n the late 1940s, Dr. Pierre Huguenard, an anesthetist at the Vugirard Hospital in Paris, was writing his medical thesis on curare, the Indian arrow poison that was being used for relaxation of muscles during abdominal operations. In scouring the literature on curare, Huguenard ran across a paper by a naval surgeon, Henri Laborit, that mentioned some effects of curare that had gone unnoticed by others.

Laborit was different in his background from other surgeons at the time, since he had studied biology, biochemistry, and pharmacology. His publications dealt mainly with the then new idea that suitable preoperative medications could diminish the degree of general anesthesia that was needed during operations and would lessen the likelihood of postoperative surgical shock, an often deadly complication.

Dr. Pierre Huguenard

In 1950, the French navy transferred Laborit from Tunisia to the Val-de-Grâce, a military hospital in Paris. Huguenard soon heard about Laborit's work, and he quickly became a believer and collaborator. Laborit was a charming and persuasive man who expressed his thoughts and opinions clearly. Huguenard was younger than Laborit, and was at times a fierce iconoclast, especially in expressing his beliefs that the anesthetist should not be looked upon as having a lesser role in the operating room. Over time, surgeons gradually came to accept the vital role that he and his fellow anesthetists played in the successful outcome of their surgical patients.

Laborit soon eliminated morphine in the original preoperative "cocktail" and settled on a combination of Dolosal and Phenergan. This combination along with ice packs caused the body temperature to fall to 33°C to 35°C. The drugs managed to dull the patients' defenses against cold, a state that Laborit called "artificial hibernation." Although this combination represented an improvement, Laborit wanted an even more active drug. He turned to the idea that other members of the antihistamine family to which Phenergan belonged might be the answer. The drug company Rhône-Poulenc had just marketed such a drug for use in parkinsonism under the trade name of Diparcol. Laborit decided to try it.

The new mixture consisting of Diparcol and Dolosal became known as "Dip-Dol." It was first used on a Madame X, who wanted to have her nose straightened. Madame X was very distressed at the prospect of undergoing any surgery. Inhalation anesthetics could not be used for the operation, since a mask placed around the nose and mouth would cover the operative site. Local anesthesia would have to be used, but the agitated Madame X was a poor candidate for any operation, especially if she remained awake during the procedure.

Faced with this situation, Huguenard suggested that he administer the new preoperative Dip-Dol mixture. He did, and Madame X was awake and calm during the entire procedure. Although her eyes were closed, she answered questions, and afterward she remarked that she was completely indifferent to her surroundings, her worries, and her perceptions. She was fully aware of the surgeon's manipulations on her nose, but she imagined that the surgery was being performed on someone else's nose. Madame X's experience was a great revelation for Huguenard. She was the first patient in whom anxiety, intense emotional instability, and fear had been subdued without loss of consciousness.

Word of this discovery quickly got around to the small group of surgeons and anesthetists in the hospital, but the hospital psychiatrists were unaware of Madame X's case. Huguenard was so impressed of the results that he presented Madame X's case to a meeting of the Paris Anesthetic

Society in 1950. But Laborit and Huguenard were still searching for a better drug than Diparcol. Laborit went back to Rhône-Poulenc looking for still other members of the antihistamine family that he could try. He discovered one lying on the shelves with the label of 4560RP, which had been synthesized a year earlier but had not been further evaluated. The compound was chemically similar to Phenergan, but it had an additional chlorine atom. The label of 4560RP was soon to be renamed chlorpromazine, and even later, Thorazine. The new combination using chlorpromazine and artificial hibernation proved to be superior to any of the previous programs that had been tried. Used alone, the drug produced a state of reversible "pharmacological lobotomy," during which the patient remained conscious but simply was totally disinterested in the surroundings.

Artificial hibernation was easily induced, and the postoperative outcomes were superior to all previous programs. Later Laborit wrote that the properties of chlorpromazine were proving to be more remarkable every day. What is outstanding about Laborit, a surgeon, is that he saw that the drug could be used for certain indications in psychiatry. It was a prophecy soon to come true.

Word of the chlorpromazine successes spread rapidly throughout the Parisian surgical and anesthetist circles, but three Val-de-Grâce psychiatrists also took note. They were the first to use the drug on a psychiatric patient. Their patient was a fifty-seven-year-old laborer who, prior to admission to the hospital, had been making polemical speeches in cafés in which he proclaimed his love of liberty as well as accosting strangers while walking down the street with a flowerpot. After a few days of treatment with chlorpromazine, his peculiar mannerisms lessened; after a week of treatment, he was telling jokes and funny stories to the medical staff. After three weeks of treatment, he was discharged in what seemed to be a normal state.

Dr. Pierre Deniker

Dr. Pierre Deniker, a psychiatrist at the Sainte-Anne Hospital in Paris, an institution for the mentally ill, realized that chlorpromazine alone might be capable of subduing highly agitated, delirious patients. At that time the conditions of such patients were dreadful. The medical community generally had treated them as outcasts with an attitude of pessimism and despair about these difficult problems. Few hospitals would admit such agitated, out-of-control patients, and those

National Library of Medicine

that would admit them would do so only for emergency treatment. With each passing year, the asylum population was growing and reaching staggering proportions.

Screaming, combative individuals were housed in overcrowded, prison-like conditions. Catatonic patients stood like statues for hours. Others lay on benches or on the floor, lost in their hallucinations. Still others were incessantly on the move, like caged animals. Shouts of raving patients day and night filled the air. Still others shrieked, spat, and threatened other patients or themselves with bodily harm. The therapies offered, psychotherapy, insulin shock, and electroshock, were generally ineffective in improving the behavior of longer-term patients.

Violent individuals were secluded, and the most dangerous ones were locked in padded rooms that had been stripped of all furniture and toilets. Others were restrained in straitjackets, bed restraints, or chains. For most, the outlook was hopeless. They were destined to remain indefinitely in the hospital, their minds firmly entrapped by their own psychoses.

Hearing from the Val-de-Grâce psychiatrists and the success with their patient, Deniker became the innovator by administering chlorpromazine to these hopeless patients without using artificial hibernation. The drug calmed the acutely agitated patients and made them indifferent to their surroundings. Straitjackets or restraints were no longer necessary. Onlookers noticed a marked decrease in the level of decibels emanating from the wards where these patients were housed. The yelling and screaming had been transformed to a welcome and unaccustomed silence. Chlorpromazine was all that was needed to calm otherwise violent patients. Deniker also found that higher doses were needed, and he increased the dose of chlorpromazine to four to six times the levels that Huguenard had used. He learned that the drug had to be given often.

Professor Jean Delay, Deniker's chief, joined Deniker in carrying out and publishing their trials of chlorpromazine. To describe the drug that selectively settles the excited mind, Delay coined the term *neuroleptic* (from *neuron*, meaning "nerve cell," and the Greek *lēptikos*, meaning "to grasp"). The word *antipsychotic*, synonymous with *neuroleptic*, became more familiar.

Later Developments

The miracle of rescuing patients from their schizophrenic hell was soon confirmed. The patients, free of their delirium, reestablished normal contact with the outside world. By the mid-1950s, the therapeutic effectiveness of chlorpromazine and the other antipsychotic drugs that followed allowed psychiatric hospitals to discharge patients at an unprecedented rate. With chlorpromazine, the hospital stay of many such patients was shortened to a week or two.

The need for frequent doses of chlorpromazine made compliance with the drug program by the discharged patients more difficult. But the shortened hospital stay was a huge change over the prior experience, when patients stayed for months, even years, receiving treatments that usually failed. It soon became apparent, however, that the acutely psychotic patients were much easier to treat than the chronically psychotic patients.

At first, Delay and Deniker's results were not accepted in many quarters of orthodox psychiatry. The psychoanalysts, especially those in the United States, rejected the results. Others were reluctant to use the drug without artificial hibernation.

Except for a few diehards, the increasing success of chlorpromazine gradually won over the psychiatric communities in Europe. Rhône-Poulenc saw a large commercial opportunity developing and marketed the drug under the name Largactil. In a trip to North America, Deniker and Huguenard extolled the virtues of the drug to medical audiences in the United States and Canada, but they were met with considerable resistance. Americans found that "Largactil" was an awkward name for the drug, and it was changed to Thorazine.

In 1954, Dr. Heinz Lehmann, clinical director of Montreal's Douglas Hospital and later chair of the McGill University department of psychiatry, is credited with bringing the practical use of chlorpromazine to the attention of the Western Hemisphere. In a single paper, Dr. Lehmann pointed out the practical significance of chlorpromazine since it had escaped serious consideration for nearly almost two years by the Canadian and American psychiatric communities.

Lehmann clearly outlined the clinical guidelines for use of the drug, described the expected results, and presented its side effects and dangers. In short, the paper was a therapeutic manual for any psychiatrist to proceed to use the drug with confidence and safety. The response in both Canada and the United States was overwhelming and confirmatory, as articles piled up waiting for an increasing time before getting published in medical journals. The application of Lehmann's recommendations was generally credited with the rapid falls in the mental institutional populations in Canada and the United States.

Comment

Schizophrenia affects 1 percent of people worldwide and is diagnosed at the average age of twenty years. The early onset, coupled with the disruptive behavior that can result from a lack of treatment, means that many patients will be on medication for the many decades of life left to them. Some patients who had been taking chlorpromazine or other similar drugs for long periods of time may develop serious side effects in the form of what is called tardive dyskinesia.

The phrase combines the word *tardive*, meaning "having a delayed onset," and *dyskinesia*, meaning literally "faulty motions," which refers to the involuntary, repetitive body movements that characterize the disorder. An afflicted patient shows purposeless, senseless movements, such as grimacing, protrusion of the tongue, puckering of lips, smacking of the mouth, and rapid blinking of the eyes. Rapid movements of the arms, legs, trunk, and fingers also occur. Such individuals are unable to voluntarily stop these movements.

There is no standard treatment for tardive dyskinesia. The first step is generally to stop or minimize the use of the drug. However, for patients with a severe underlying condition, this may not be a feasible option. Replacing the drug with substitute drugs may help some patients. Symptoms of tardive dyskinesia may remain long after discontinuation of antipsychotic drugs, but with careful management, some symptoms may improve and/ or disappear with time.

Honors

In 1957, the Albert Lasker Foundation awarded its highly respected prize for clinical research to Drs. Pierre Deniker, Henri Laborit, and Heinz Lehmann for their work in bringing chlorpromazine to the forefront of psychiatric treatments. The following includes excerpts from the Lasker citations of each honoree.

Dr. Pierre Deniker's citation was "for his introduction of chlorpromazine into psychiatry, and for his demonstration that a medication can influence the clinical course of the major psychoses." In doing so he launched a whole new method of psychiatric treatment and research. His work also spawned the developing field of laboratory research for the study of the biochemistry of mental disorders.

Dr. Henri Laborit was honored "for his extensive studies of surgical shock and post-operative illness which resulted in the first application of chlorpromazine as a therapeutic agent." Although his discovery led to an improvement in the management of surgical problems, even more importantly, he recognized the implications of his discovery for clinical psychiatry. He is often credited with bringing psychiatry into the medical mainstream where heretofore it had been treated as a curious orphan.

Dr. Heinz Lehmann's citation was "for his demonstrations of the clinical uses of chlorpromazine in the treatment of mental disorders." In a single published paper, he was responsible for effectively presenting the use of chlorpromazine in the United States and Canada by setting forth practical clinical guidelines and presenting its dangers. His work found rapid and wide application.

Chance played a role in several events in the discovery of chlorpromazine. In persisting to find a better drug for postoperative patient care, Henri Laborit discovered chlorpromazine, which was idly collecting dust on the shelves of Rhône-Poulenc and labeled simply as 4560RP. There were a number of other antihistamines from which he might have selected, but he chose this one, because its structure closely resembled Phenergan except for its having a chlorine atom.

But once he saw what 4560RP could do for his patients, he spread the word of his success. Beyond that, he saw that chlorpromazine might be useful for treating mental illnesses. In making the connection, Laborit's fertile mind could see far beyond the boundaries of surgery to suggest that chlorpromazine was potentially useful in psychiatry, a distant specialty that was at the time not yet fully recognized as a legitimate field of medicine.

REFERENCES

Baldessarini, R. et al. "Pharmacotherapy of Psychosis and Mania," in *The Pharmacological Basis of Therapeutics*, 11th ed., *by L. Goodman and A. Gilman*. New York: McGraw-Hill, 2005.

Healy, D. *The Creation of Psychopharmacology*. Cambridge, MA: Harvard University Press, 2004.

Stahl, S. M. *Psychopharmacology of Antidepressants*. London: Dunitz, 1997.

Thuillier, J. *Ten Years that Changed the Face of Mental Illness*. Cambridge, MA: Blackwell Science, 1999.

Part VI
The Vitamins (vit(al) amines)

20

The Naming of Vitamins: The Vital Missing Pieces

The word, "Vitamine," was a catchword, but it was not chosen by mere accident.

—*Casmir Funk, 1911*

Introduction

For the first twenty or so years of the twentieth century, an important field of research concentrated on the measurements of human energy metabolism under various conditions. These results found a direct application to human nutrition since they directly related to the amounts and distributions of carbohydrate, protein, and fat in the energy mix of the diet. Somewhat later, salts and minerals were recognized as being important and were added to what was then believed to be a complete diet.

Prior to 1900, the diet of animals was of little interest. Animals fed a so-called purified diet died when one of the major nutrients was missing, and that was clear enough proof that the diet had to be complete with the major nutrients, including salts. But a few studies in animals during this period revealed that the "complete diet" was not complete. Rather it lacked an unknown "something or somethings" that prevented normal growth of animals.

An early pioneer in vitamin research was Christiaan Eijkman. In 1886; Eijkman observed that chickens fed white rice developed a neurological disorder that could be cured by feeding brown rice. It was the bran of the rice that contained the unknown factor. In 1911, a Polish chemist, Casmir Funk, isolated the active factor and found that it was an amine. He christened this substance a "vitamine," by combining *vit* from *vital* with *amine*, to form the word *vitamine*. Funk prophesied that other "vitamines," yet to be discovered, would also be amines. Professor Elmer McCollum, a pioneer in nutritional research, opposed adopting Funk's coined word *vitamine* because he thought a vitamine was no more "vital" than any other nutrient and because he thought that not all vitamines would prove to be amines, which proved the case when new research demonstrated that other "vitamines" lacked an amine group. The word was shortened to *vitamin*, the name changed to its current spelling in 1920.

In 1907, two Norwegian scientists, Axel Holst and Theodor Fröhlich, speculated that a disease that afflicted the crews of Norwegian fishing fleets called "shipboard beriberi" bore a strong resemblance to scurvy. Fortuitously they selected the guinea pig as their animal model, unaware of the fact that the guinea pig is one of a very few mammals that cannot make ascorbic acid. When they fed guinea pigs a diet consisting of various types of whole grain foods, they developed a scurvy-like disease that could be prevented by supplements of fresh cabbage or lemon juice. This anti-scurvy factor, chemically ascorbic acid, later became known vitamin C.

Meanwhile, Professor Elmer McCollum in the United States was making important advances using animals fed different diets. He found that rats fed a purified diet began to lose weight after ten weeks, but recovered when small amounts of a purified fat-soluble fraction of butterfat was fed. He named this factor A, which became vitamin A. He also showed that rickets in animals could be treated with a fat-soluble vitamin that was separate from vitamin A, and it became vitamin D. McCollum and Marguerite Davis gave the "factors" letter names, because their structures had not yet been determined to give them proper chemical names. McCollum's experiments also led to the discovery of the water-soluble vitamin B, which he later showed was not a single compound, but a complex of other yet-to-be-discovered water-soluble vitamins. He proposed that the first of these was the factor in rice husks and named it vitamin B1. As the list of new discoveries grew, a system of nomenclature was developed in which they had letter names before their structures had been determined to give them proper chemical names. The preponderance of this work led to the development of a definition of a vitamin as a tiny amount of an organic substance that is not made in the body that is essential for the health of an animal or man.

During the first decade of the twentieth century, Professor Frederick Gowland Hopkins (later Sir Gowland and a Nobel laureate, 1929) conducted experiments whose results could be explained only by a basic deficiency of an unidentified factor or factors in the basic diet, which he called an "accessory food factor" and would eventually be called a vitamin. Reflecting on his experiments, Hopkins prophesied that "no animal can live on a mixture of only carbohydrate, protein, and fats with supplementary minerals. . . . The animal body is adjusted to live either on plant tissues or on other animals, and these contain countless substances other than protein, carbohydrate, and fat. Physiological evolution, I believe, has made some of these well-nigh as essential as the basal constituents of the diet. The field is almost unexplored. In diseases such as rickets and in particular scurvy, we have had knowledge of a dietetic factor, but though we know how to benefit these conditions empirically, the real errors of the diet are to this day quite obscure."

This conclusion was reemphasized by Funk in 1912: that beriberi, scurvy, pellagra, and rickets were each cause by the deficiency of a specific special substance. Unlike Hopkins, Funk published his paper in a medical journal, and it was widely quoted. That was an important step toward acceptance of the idea of a deficiency disease. Even though the dietary origin of rickets and scurvy and the ability to treat these disorders by dietary means were becoming more clearly defined, many medical men had to be convinced of the then foreign concept that a lack of something in the diet could cause a disease.

The Vitamin Alphabet

At present, there are thirteen well-established substances that fit the earlier definition. Given this situation, the nomenclature became rather confusing. The following is an attempt to clarify the present state of naming the vitamins. They fall into two large classes, namely those are soluble in fat and oil and those that are soluble in water. In addition to having a name, each vitamin has a shorter chemical name that is used to describe it. The fat-soluble vitamins are as follows: vitamin A, retinol; vitamin D, cholecalciferol; vitamin E, tocopherol; and vitamin K, naphthoquinone. Vitamins B and C complete the sequence of A to D; they include vitamin B, now a large subclass of water-soluble vitamins, and vitamin C (ascorbic acid), which is water soluble.

The alphabetical intervening sequence following vitamin E is vitamins F, G, H, I, and J. Vitamin F turned out to be the essential fatty acids that were already recognized elsewhere. Vitamin G was vitamin B_2, riboflavin. Vitamin H was vitamin B_7, biotin. Vitamin I was a nickname for ibuprofen, a drug, and Vitamin J was a catechol that is normally produced by the body. The rest of the alphabetic sequence beyond K lacks any legitimate members. Thus vitamin L, anthranilic acid, was synthesized by the body; vitamin M was folic acid, which is vitamin B_9. The final sequence of vitamins

O, P, PP, S, and U are compounds that are either synthesized in the body or are normal metabolic breakdown products produced by the body.

The water-soluble class consists of the large subclass of vitamin B, which is now represented by eight members, which include vitamin B_1, thiamin; vitamin B_2, riboflavin; vitamin B_3, niacin; vitamin B_5, pantothenic acid; vitamin B_6, pyridoxine; vitamin B_7, biotin; vitamin B_9, folic acid; and vitamin B_{12}, cobalamin. The missing numbers, B_4, B_8, .B_{10}, and B_{11}, were initially thought to be vitamins, but later work showed that they did not satisfy the definition of a vitamin. Vitamin C is formally an "alphabetic vitamin," but chemically it belongs in the water-soluble class.

Comment

The preceding discussion is a small testament to the very large amount of vitamin research that occurred during the twentieth century and the first decades of the twenty-first. During this period vitamin research has been honored by no fewer than seven Nobel Prize winners: five in Physiology or Medicine and two in Chemistry. The winners were Adolf Windaus in 1928 (vitamin D); Christiaan Eijkman in 1929 (vitamin B_1); Frederick G. Hopkins in 1929 (accessory food factors); Paul Karrer in 1937 (vitamin K); Albert Szent-Györgyi in 1937 (vitamin C); Richard Kuhn in 1938 (Vitamins B_2 and B_6); Henrik Dam in 1943 (vitamin K); and Edward Doisy in 1943 (vitamin K).

REFERENCES

Kutsky, R. J. *Handbook Of Vitamins and Hormones.* New York: Van Nostrand Reinhold, 1973.

The Merck Manual. 1997 Nutritional Disorders; Vitamins and Minerals, Chapt 135, pp 652–660. Whitehouse Station, N.J. Merck Research Laboratories.

USDA Table of Nutrient Retention Factors. Release. *USDA.* December 2007.

Wikipedia, the Free Encyclopedia. "Vitamin."

C H A P T E R

21

Vitamin B₁: Brown Rice Cures Beriberi

A vitamin is a substance that makes you ill if you don't eat it.
—*Albert Szent-Györgyi, chemist*

Beriberi

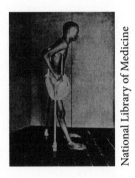

National Library of Medicine

Beginning in 1602 and extending over the next two centuries, the Dutch East India Company had a monopoly on the rich trade of the so-called Maluku, often referred to as the Spice Islands, by successfully subduing the local rulers and by driving out both the Portuguese and the British from Malaysia and Indonesia. In the Indies, the company encountered new diseases in their employees and in the Dutch army. To deal with these disorders, the Dutch government sent teams of physicians out to Batavia (now Djakarta) to investigate and treat these often-fatal maladies.

Jacobs Bontius, a Dutch physician, entered the service of the Dutch East India Company in 1627, and during the four years he spent in the Spice Islands, he immersed himself in the study of tropical diseases. He published a book, *De Medicina Indorum*, in 1642. In the book he described a native sickness called beriberi (a duplication of a Singhalese word, *beri*,

Vitamin B_1: Brown Rice Cures Beriberi

meaning "weakness"). It was the first European description of the disease: "I believe that those, whom this disease attacks, with their knees shaking, and their legs raised up, walk like sheep. It is a kind of paralysis . . . for it penetrates the motion and sensation of the hands and feet and sometimes the whole body." Bontius had no idea of the cause of the disease, nor did he discover any effective treatment. Beriberi raged on for the next two centuries throughout the world. The disease caused progressive weakness, loss of sensation in the legs, and mental disorders that became known as "dry beriberi." A second form of the disease called "wet beriberi" caused heart failure with generalized edema.

But it was not until the end of the seventeenth and the beginning of the eighteenth century that the disease attracted more general attention. Subsequently, it made its appearance on five continents, but countries in Eastern and Southeastern Asia were hardest hit. In 1871 and 1879, Tokyo suffered widespread epidemics that became known as the "national disease of Japan." The Japanese navy, especially when at sea, was most severely affected. In 1882, a Japanese training ship set out on a 272-day voyage from Japan to Honolulu via New Zealand and Chile. Of the 275 men aboard, 169 developed beriberi and 25 died. Admiral Kanehiro Takaki, a naval surgeon alarmed at the devastation inflicted on the sailors, suggested that the low protein content of the seafaring diet was the culprit. On the next voyage, he ordered an increased protein intake and a replacement of barley with rice. The incidence of beriberi fell to 3 percent. Takaki thought that the protein intake was the obvious answer; he had entirely overlooked the possible role of the rice.

Beginning in the mid-nineteenth century, the Dutch and the English were again at war, each trying to wrest control of Sumatra, the largest island in what is now Indonesia. The Dutch were more powerful, and the English eventually withdrew. The northern part of the island was the home of a war-like people, the Achinese. They bitterly resented the Dutch and organized a well-armed resistance to protect their homeland. As the fighting dragged on, Dutch soldiers developed beriberi, and thousands of the troops were lost. Yet the Achinese were completely unaffected by the disease. At the time no one thought that the foods eaten by the two belligerents had anything to do with beriberi.

Medical thought of the day was heavily influenced by Louis Pasteur's successes in focusing on microbes as causes of disease. Infections and toxins were the only two accepted causes of illness. Both in a sense were positive external factors exerted on the body. It was a completely foreign concept that a negative factor, such as a lack of something, could cause a disease.

Christiaan Eijkman

Christiaan Eijkman, a Hollander, was an exceptionally bright graduate of the Dutch Military Medical School, following which he worked in a physiology laboratory for two years. In 1883, Eijkman left Holland for the Indies as a medical officer for health in villages in Java and Sumatra. Two years later, having contracted malaria, he returned to Amsterdam. Following recuperation, he moved to Berlin and worked in Robert Koch's bacteriological laboratory. By chance, Professor C. A. Pekelharing was visiting Koch's laboratory prior to departing for the Indies as the leader of the newly formed Royal Commission on Beriberi, which the government had established in response to the devastating losses being sustained by the Dutch troops in Sumatra. Professor Pekelharing, impressed with Eijkman's intelligence, offered him the job as an assistant on the commission, which he promptly accepted.

After nine months of investigation, the commission members returned to Holland in 1887 to report that although beriberi certainly had an infectious origin, their search for the offending microbe had been unsuccessful. The commission's opinion was strongly influenced by the times, and especially by the ongoing discoveries of various bacteria that were responsible for no fewer than twenty-two diseases during the decade of the 1880s. It was natural, therefore, that this commission should find it easy to believe beriberi was an infection of some sort, and therefore the main problem in finding its cause was to identify the organism. It followed that the proper treatment was disinfection of living quarters and related measures.

The commission also recommended that further work was required for confirmation of this idea, and they recommended that Christiaan Eijkman be appointed director of Geneeskundig Laboratorium (medical laboratory) in the military hospital in Batavia that was caring for many patients with beriberi. As the director of the hospital, Eijkman continued the hunt for the elusive germ. He served as director for the period of 1888 to 1896, and during that time he made important studies showing that European people living in tropical regions maintained perfectly normal physiological parameters such as blood counts, respiratory functions, perspiration, and the like. But he didn't find the infectious cause of beriberi.

The Discovery

Eijkman did make an important discovery that occurred by chance. Glancing out at the courtyard of the hospital, Eijkman noticed that the chickens there had contracted a bizarre form of paralysis. Over the next days, forty-one of forty-seven hens were affected, and thirty had died.

Eijkman thought that the progressive paralysis resembled the neuritis of human "dry beriberi." Suddenly, the epidemic of the chickens was over, and the chickens were again healthy. Eijkman learned that the cook used leftover white rice from the hospital as chicken feed, but the cook's successor objected to military rice going to chickens, and he switched the feed back to brown unhulled rice.

Eijkman repeated the white rice/brown rice experiment several times, always with the same results. Clearly the brown rice contained something that cured the white rice–induced neuritis. To find an infectious cause of beriberi, he first tried unsuccessfully to transmit the disease from a sick hen to a healthy one. And even later, he failed to transmit the disease from a patient with beriberi to a normal hen. Eijkman then took the next step and found that feeding the coatings of the brown rice cured paralyzed chickens in four hours. Despite the ridicule of Eijkman's colleagues that a hen could develop beriberi, he stubbornly persisted in trying to solve the mystery. But after viewing the curative effects of feeding the rice coatings, he abandoned any further search for a microbe.

The Response

Following the orthodox medical thought of the day, Eijkman concluded that the white rice contained a toxin, and the coatings of the brown rice contained the antidote. At the time, a number of theoretical causes, all related to a poison in the white rice, were proposed. One strong advocate of the toxin theory was Dr. Evart Van Dieren, a practicing physician in Amsterdam. Van Dieren based his conclusions entirely on a review of the literature, but his lack of experience with clinical cases of beriberi damaged his reputation as an expert. Sir Patrick Manson speculated that the disease was due to a toxin elaborated by a germ that grew outside the body and that improper handling of rice led to its infection. Sir Ronald Ross believed that beriberi was due to arsenic poisoning, while still others held that beriberi was due to lack of nutrients in the diet.

Eijkman published his results in a Dutch medical journal in 1890. His paper was the first to be based on direct experiments. Many in the Dutch orthodox scientific establishment met the paper with heavy criticisms. Because of the language barrier, the paper received little attention outside the Netherlands. But Admiral Takaki in Japan eventually learned of Eijkman's findings and confirmed them in his own experiments.

During his six-year stay in Batavia, Eijkman firmly established that the neuritis in birds was caused by a diet that consisted exclusively of polished rice and that it could be prevented or cured by feeding unpolished rice. Although he discarded the toxin theory, Eijkman still hung onto a vague infectious theory, but indirectly he realized that something in the white rice diet and in the other foods was missing, although he never came out and explicitly said so. He was sharply criticized by doctors and scientists in Holland who failed to believe that an unsteady gait in hens was analogous to the human disease of beriberi. They also complained that his long stay in Batavia had yielded such a meager amount of new information.

Gerrit Grijns

In 1896, Eijkman, plagued by continuing ill health, returned to Holland and was replaced by a young colleague, Gerrit Grijns. Grijns had a scientific mind, and he addressed the unanswered questions raised by his predecessor's work. He believed Eijkman's results made the rice toxin-antidote theory untenable. Step by step, he was able to exclude various other possibilities, particularly the idea that the presence of starch was essential for the causation of neuritis. He eliminated a lack of salt, fat, or protein as causes of the disease. Feeding the birds exclusively on meat that had been autoclaved at 120°C produced the disease. He also discovered that the protective substance in rice bran was present in other articles of food, notably the legume known as katjang idjoe (native beans).

All of these results made the rice toxin-antidote theory untenable. After three years of work, Grijns concluded that the neuritis in chickens, and by analogy to human beriberi, was due to a deficiency of a certain but still unknown factor. He published a landmark statement in 1901, which was the first to definitely state that there is a direct relationship between a dietary component and a disease. In his own words: "There occurs in natural foods, substances, which cannot be absent without serious injury to the peripheral nervous system. The distribution of these substances in different foodstuffs is very wide. The separation of these substances meets with the difficulty that they are so easily disintegrated. They cannot be replaced by simple compounds."

Grijns's work led to the tests on insane inmates by Hulshoff Pol, an Indonesian physician. In the first trials on thirty patients with longstanding beriberi, feeding katjang idjoe daily produced decided improvement, but the paralysis was slowly reversible, if at all. In another trial, Pol tested the preventative effects of this legume in three groups of inmates, all of whom were free of beriberi at the onset of the study. One group that received 150 grams of katjang idjoe daily in place of one of the rice meals remained free of beriberi. A second group that received raw or partially cooked vegetables other than the beans yielded inconclusive results. Most of the inmates in the third group who ate the customary white rice developed overt beriberi.

This experiment demonstrated that beriberi must be due to some deficiency in the diet, since, as E. B. Vedder commented, "It is hardly conceivable that if the rice were toxic that the disease could be prevented simply by the addition of another vegetable to the diet. If we administer an amount of arsenic or other known poison sufficient to produce intoxication, we do not expect that these effects would be counteracted in the least by the administration of beans or any other food."

Other Human Studies

Even though the idea that chickens could develop dry beriberi seemed far-fetched, Eijkman, prior to returning to Holland, had convinced Adolphe Vorderman, the sanitary inspector of Java, that rice feedings were the key to the disease. Vorderman was in charge of the health conditions of the 101 prisons in Java, and he set out to study the incidence of beriberi in the inmates in these facilities. Beriberi was so rampant in the prisons that as short a stay as three months was a death sentence. Deterioration of the prison buildings and poor sanitary conditions were thought to be main reasons for the high incidence of the disease, perhaps through transmitting an infection or through the ingestion of a toxin.

Vorderman found only nine cases of beriberi in the 100,000 prisoners fed unpolished rice. But 4,000 cases occurred in the 150,000 prisoners fed white rice. These results seemed clear enough, particularly since the living conditions of the various prisoners showed no relationship to contracting the disease. Vorderman published his report in 1897, but unfortunately it was written in Dutch and therefore reached a limited audience. In his report, Vorderman, still under the spell of traditional teachings, concluded that beriberi was caused by an infection and that the type of rice favorably or unfavorable influenced its course. Van Dieren, the champion of the toxin theory, aggressively attacked Vorderman's methods and quibbled over his statistics. Van Dieren's writings, published in two books and numerous pamphlets, seriously undermined Vorderman's work so that few medical people in Holland regarded it as a serious contribution.

In 1905, Dr. William Fletcher, unaware of the results of Vorderman, Grijns, and Pol, carried out a decisive experiment in an insane asylum in Kuala Lumpur. Fletcher devised a simple but ethically brutal study of the disease, which he described as follows: "All the lunatics were drawn up in the dining shed and numbered off from the left. The odd numbers were domiciled in the ward on the east side and no alteration was made in their (white rice) diet. The even numbers were quartered in the ward on the west side and received the same ration as the others with the exception that they were given brown rice instead of white rice. The two groups of patients were kept in separate wards, fed at different times of the day with food that had been prepared separately. Otherwise, the patients were allowed to associate with each other."

Of the one hundred patients fed white rice, fifty-three developed beriberi, but no cases occurred in the one hundred who ate brown rice. Fletcher concluded that a toxin in the white rice might cause beriberi due to a deficiency of protein intake or that white rice did not form a sufficiently nutritive diet and therefore rendered the patients' systems susceptible to an invasion by a specific microbe. Fletcher was getting close to the idea of a deficiency disease, but neither he nor any of his predecessors were able to completely discard some type of infection or toxin as a cause of beriberi.

In the laboratory, Henry Fraser and A. T. Stanton confirmed that extracts of brown rice cured the paralysis of white rice–fed chickens. In their paper presented before the Far Eastern Association of Tropical Medicine in 1910, they confirmed again that white rice produces beriberi and brown rice prevents it, and they added "that white rice makes default of some substance or substances essential for the normal metabolism of nerve tissues." It was the first time that a deficiency of something might be a cause of a disease! Since dry beriberi attacked the nervous system, the protective agent became known as an "anti-neuritic factor."

In 1906, Sir Frederick Gowland Hopkins, a highly respected biochemical experimentalist, made the groundbreaking statement that "accessory food factors," by which he meant factors present in only very small amounts in the diet, were essential for the body and that their absence could lead to abnormal states.

Despite Hopkins's stature, there was a great reluctance to accept his idea that a lack of something could cause a disease. Even as late as 1919, other experts issued a report under the aegis of the British Medical Research Council that "disease is so generally associated with positive agents, such as a parasite or a toxin, that the thought of the pathologist turns naturally to such positive associations and seems to believe with difficulty in causation prefixed by a minus sign." Yet it was becoming clear to the world that the absence of a small amount of a substance in the diet caused disease. What was missing was the chemical identity of such a substance. The chemical nature of the anti-neuritic factor in rice skins was the place to start.

Identifying the Chemistry of Thiamin

Casmir Funk, a brilliant Polish scientist working in the Lister Institute in London, took up the challenge in 1911. Starting with nearly 400 kilograms of rice polishings, he extracted an impure mixture, 200 milligrams of which cured pigeons with beriberi in three hours. Although Funk was unsure of the precise chemistry of this factor, he thought it was an amine, and he named it vital-amine, later shortened to vitamine and still later to vitamin. Eventually, the anti-beriberi vitamin was named thiamine, shortened to thiamin, and as other vitamins were discovered, it became known as vitamin B_1 in recognition of it being the first vitamin to be discovered. Thiamin's structure is unusual in that it contains sulfur. The name thiamine recognized the sulfur as *thi*, shortened from the accepted prefix for sulfur-containing chemicals.

In 1926, thiamin was the first vitamin to be obtained in crystalline form, and the quest for determining its exact chemical nature was led by a chemist, Dr. Robert R. Williams. As a young man, Williams had been stationed in the Bureau of Science in Manila. In 1910, E. B. Vedder, an army captain and an avid student of beriberi, presented Williams with a quart of brown liquid that contained an extract of brown rice, one milliliter of which would cure the beriberi-induced neuritis of chickens. Vedder challenged Williams to conduct a detailed analysis of the extract so that each separate component of the mixture could be tested for its activity in the chickens. It was a formidable request, and it took the perseverance of Williams over the next thirty years to unravel the precise chemistry of thiamin and to synthesize it.

During the 1920s, thiamin was touted as a miracle medicine. The margin of safety was wide, and physicians prescribed large doses of the vitamin for otherwise normal individuals suffering from "subclinical deficient states." Thiamin was recommended as a "cure" for anorexia, weight loss, neuritis, neuroses, heart failure, and gout. Its popularity faded rapidly when the hoped-for cures for these diseases failed to materialize. But thanks to the fortification of rice with the vitamin, beriberi has disappeared in many parts of the world. White wheat flour and many other foods are now fortified with a powdered mix of vitamins and minerals.

Recognition

Christiaan Eijkman and Sir Frederick Gowland Hopkins shared the 1929 Nobel Prize in Physiology or Medicine. The Nobel Committee had first entertained Eijkman a decade earlier for his discovery, but they felt that his work was already too old, since it had been completed nearly thirty years before. Yet in 1929, the committee reconsidered, opining that the age of the discovery was not relevant if the discoverer had opened up a new field. Hopkins's seminal experiments that were the first to define deficiency states in animals were also old, having been made in 1905, but they too had opened the whole field of nutritional deficiency diseases. Eijkman's citation of the award was "for the discovery of the anti-neuritic vitamin." Hopkins was cited "for his discovery of the growth-stimulating factors." Eijkman was too ill to travel to Stockholm to receive the prize in person.

Comment

Christiaan Eijkman was the first to clearly define the existence of what Hopkins later called an accessory food factor. At the time of Eijkman's discovery, traditional medical thought held that a complete healthy diet consisted of carbohydrate, protein, and fat plus adequate amounts of salts and minerals. All diseases were caused either by external delivered infections or by toxins. The idea that white rice could cause a disease and that the hulls of brown rice could cure it was so foreign to the prevailing mindset that it simply could not be accepted.

Only with time and more definitive studies, plus the work of Casmir Funk in isolating the anti-neuritic factor, did the medical public come to realize that a deficiency of a tiny amount of a factor in food could produce such a devastating disease as beriberi. The problem of changing a well-established mindset of medical thought is a recurrent one that is seen in many other discoveries that are presented in this book.

Eijkman's discovery joins the roles of many other great discoveries that began with an element of chance. He made the chance observation of the paralyzed hens, but his mind, prepared by his thoughts about beriberi, saw more than an isolated occurrence. Rather he saw its possible implications, and he launched a whole new pathway of inquiry, which in the end yielded spectacularly important results.

REFERENCES

Kutsky, R. J. Handbook Of Vitamins and Hormones 1973 New York, Van Nostrand Reinhold.

The Merck Manual. 1997 Nutritional Disorders; Vitamins and Minerals, Chapt 135, pp 652–660. Whitehouse Station, N.J. Merck Research Laboratoriies.

USDA Table of Nutrient Retention Factors. Release 6.*USDA*. December 2007.

Wikipedia, the free encyclopedia: Vitamin.

22

Vitamin C: Another Battle for Priority

The most sudden and visible good effects were perceived from the use of oranges and lemons.

—James Lind, 1753

Curing Scurvy

Scurvy is an ancient disease that affected people whose diets lacked an abundant supply of raw fruits and vegetables. Although the disease was often rampant in soldiers and sailors, it seldom attacked the officers. Babies were very susceptible to the ravages of scurvy. The disease causes bleeding, infected gums, and hemorrhages in the skin and bones, resulting in weakness and eventually in coma and death.

In 1885, Professor August Hirsch devoted sixty-one pages to scurvy in his *Handbook of Geographical and Historical Pathology*. Even at that time, there was already a sizable amount of medical literature, as witnessed by Hirsch's citation of 178 relevant papers. An important citation was *A Treatise on the Scurvy* by James Lind, in which he related his results in treating scurvy. In May 1747, Lind studied twelve seamen with scurvy on board the HMS *Salisbury*. Each suffered from putrid gums, hemorrhages in the skin, lassitude, and weakness of the legs. Lind divided the twelve into pairs, and each pair received a different diet. One pair received two oranges and a lemon for six days until the supply ran out. Other pairs were given different foods without any citrus fruits. Dramatic changes occurred only in the first pair, one of whom quickly recovered enough to return to duty, while the other eventually returned to health but more slowly. In this simple experiment, Lind produced convincing evidence that citrus fruits were a valuable anti-scurvy remedy.

The problem with this simple tale is that Lind, in the 358 pages of his *Treatise on the Scurvy*, never fully embraced the anti-scurvy effects of lemons and oranges as the one definite and dramatic measure that could prevent or cure scurvy aboard ships. He wrote about the unhealthy conditions of life at sea as contributing factors. He believed that life at sea closed the pores of the skin, thus interfering with adequate perspiration, which he regarded as the principal means of the body to dispose of waste products. Even during the war with France, Lind treated hundreds of scorbutic patients with lemons with good results, but he never firmly grasped the idea that lemons were the one sure cure for scurvy.

A certain "Mr. Young of the Navy," of whom nothing more is known, expanded Lind's findings. In 1782, Charles Curtis, a ship's surgeon, related Mr. Young's plan: "Nothing more is necessary for the cure of this disease but a fresh vegetable diet of greens or roots in sufficient quantity. To be sure, we cannot have a kitchen garden at sea, but beans and peas and barley and other seeds brought under the vegetating process are converted into a growing plant and if eaten in this state without any sort of preparation is the cure of scurvy." We are uncertain as to whether Mr. Young's plans were ever fully implemented, but he was certainly on the right track.

The next big step in unveiling the mysteries of scurvy occurred during the first decade of the twentieth century. Two Norwegian experimentalists, Holst and Fröhlich, showed that guinea pigs fed cereal grains died of scurvy in twenty to forty days unless their intake was supplemented with raw green vegetables. But these anti-scurvy foods lost their effectiveness when heated to 100°C. Strangely enough, these important results, first made in 1907, were universally ignored as being inapplicable to human scurvy. But one of the most significant findings of Hoist and Fröhlich was the production of scurvy in the guinea pig, a finding that would be key to much of the future research. Indeed, the guinea pig is the only laboratory animal that requires the anti-scurvy factor, and it became the animal that found wide usage in the experimental research that surged ahead, so that by 1920, the anti-scurvy agent was finally recognized as an essential nutrient for human nutrition. It was renamed vitamin C, recognizing that by that time only vitamins A and B were known.

Unfortunately, the human application of experimental results with scurvy seriously lagged behind because of uncertain results. Pediatricians at the time were very concerned about developing measures for the prevention and treatment of scurvy in infants and children. In 1898, infantile scurvy was so common that the American Academy of Pediatrics convened a special committee to assess the incidence of scurvy in relation to dietary intake. The study involved 372 infants. Only ten of these were fed breast milk, and none of the ten had scurvy. The remaining infants all received heated, milk-based formulas, and many developed scurvy. When fruit juices were added to the intake, scurvy was promptly cured. The medical community finally realized that heat destroyed the anti-scurvy factor. It was no accident that the upsurge in the incidence of scurvy in infants and children beginning in the 1890s coincided with the increasing use of both heat-sterilized artificial formulas and pasteurized milk. By the turn of the century, measures for the prevention and treatment of scurvy were understood, but the mystery of the chemical nature of the anti-scurvy factor remained elusive.

Szent-Györgyi: Maverick Chemist

Public domain

Albert Imre Szent-Györgyi von Nagyropolt was born in 1893, the scion of a titled Hungarian family. For his many friends in Western Europe and in America, he simplified his name, preferring to be called "Saint Georgie," and he abandoned the "von" as being "just plain stupid." As a young man, Albert announced to his family that he intended to become a medical researcher, but his uncle, anatomist Mihály Lenhossék, strenuously discouraged the idea. Science, he said, had no place for "dimwits" such as Albert. He suggested that his nephew pursue a career in cosmetics, dentistry, or pharmacy, but never science! Lenhossék relented when Szent-Györgyi graduated high school with honors. Albert entered the Budapest Medical School in 1911. Soon bored with his medical courses, he gravitated to his uncle's anatomy laboratory. Lenhossék allowed him to work there with one condition: Albert's first research would focus on the anatomy of the human rectum and anus. (Mihály apparently suffered from hemorrhoids and hoped to profit from his nephew's investigations.) In fact, Szent-Györgyi's first scientific article, published in 1913, dealt with the epithelium of the anus. "Because of my uncle," he often joked later, "I started science at the wrong end."

World War I interrupted Szent-Györgyi's medical education. In the summer of 1914, he began serving as an army medic in the Austro-Hungarian Army. Though he earned a medal of valor for his bravery, after two years in the trenches, he was disgusted with the war and despaired of ever surviving it. He carefully shot himself through the left humerus, claimed he had been hit by enemy fire, and was sent back to Budapest. While his arm was healing, Szent-Györgyi finished medical school and received his MD degree in 1917. Later that year he married Cornelia ("Nelly") Demeny. Their only child, Cornelia ("Little Nelly"), was born in October 1918, just before the war ended.

After receiving his MD degree, Szent-Györgyi abandoned all clinical work, preferring to pursue basic chemical research instead. He was intrigued by the bronze color of the skin of patients with Addison's disease, which is due to a deficiency of the hormones of the adrenal glands. On a hunch, Szent-Györgyi thought that this discoloration was related to the browning that is seen when certain plant foods, such as apples, are cut and exposed to the air, while others such as lemons brown only after a considerable delay. Using a simple test, he showed that the browning was due to an enzyme present in large amounts in some plants, but was only weakly active in citrus fruits. This antioxidant effect was due to a chemical that he determined to be a reducing substance.

Following his lifelong credo that "I see what others see, but think what no one else has thought," he began a search for the reducing substance in adrenal glands, believing that it might be a new hormone. He injected an extract of the gland into a single stray cat whose adrenals had been removed. The dying cat responded by "jumping off the table." Based on this single experiment, Szent-Györgyi persuaded Henry Dale (later Sir Henry) to accept him for a three-month fellowship, during which he worked feverishly preparing extracts from ground-up cow adrenal tissue. He was delighted that these extracts showed marked reducing properties in his test system. Only later did he discover that the positive test reactions were the result of tiny iron contaminants from the meat grinder and not from the adrenal extracts.

For the next several years, Szent-Györgyi and his little family lurched from one temporary laboratory position to another, always on the brink of financial disaster. Even so, he managed to publish a few papers dealing with plant metabolism. Stranded and unemployed in Hamburg and nearly starving, Albert sent Nelly and Little Nelly to Budapest. Alone and desperate, he seriously toyed with suicide. In his despair, he attended a scientific meeting at which Professor Frederick Gowland Hopkins (later Sir Gowland and a Nobel laureate, 1929) was a featured speaker. In his remarks, Hopkins made two complimentary references to a paper of

Szent-Györgyi's that dealt with oxidation of plants. After the speech, Szent-Györgyi introduced himself to Hopkins, who offered him a junior position in his laboratory at Cambridge, which also included financial support from the Rockefeller Foundation.

Discovery of the Reducing Substance

The financial pressures were now relieved, and Szent-Györgyi enjoyed working under the tutelage of Hopkins as well as joining the collegial Cambridge scientific community. Meanwhile, he continued his search for identifying the elusive reducing substance in plants. He finally isolated a reducing substance from orange juice, and he was able to determine chemically that it was some type of a six-carbon sugar acid. Hopkins urged him to publish the results, and Szent-Györgyi prepared a paper for submission to the prestigious *Biochemical Journal*.

Since he had no clear idea of the detailed chemistry of the substance, he coined the name of Ignose, a combination of a Latin root from *ignoseo*, meaning "I don't know," and the suffix *-ose* as the accepted chemical ending for a sugar. The editor of the journal did not appreciate this attempt at humor and sent the paper back asking for a new name. This time Szent-Györgyi proposed "Godnose." The exasperated editor rejected this second attempt and named the substance hexuronic acid. What Szent-Györgyi didn't know at the time was that his hexuronic acid crystals would eventually be named vitamin C.

In 1928, the Rockefeller Foundation awarded a large grant to stimulate the development of a major medical research center in the provincial Hungarian town of Szeged, where Szent-Györgyi would become the professor of the new department of medicinal chemistry. While waiting for the new laboratories to be built, he accepted the invitation from the Mayo Clinic in Rochester, Minnesota, to renew his search for the reducing substance in the adrenal glands. The Mayo Clinic was especially attractive, since Dr. Edward C. Kendall, a leading expert on the chemistry of the adrenal hormones, was working there.

The environment and facilities at Mayo were the best Szent-Györgyi had ever seen. In a surprisingly short time, he had isolated and crystallized thirty milligrams of hexuronic acid, having started with several thousand pounds of adrenal glands that were available from the stockyards in neighboring St. Paul. He sent half of the crystals to English chemist Dr. Norman Haworth, the world's foremost authority on the chemistry of sugars, for more detailed analytical work on the substance. Szent-Györgyi hoped that this was an adrenal hormone, but he was disappointed to find that it was ineffective in treating animals whose adrenal glands had been removed.

The Battle for Priority

Thus far, there had been little thought about the relationship of hexuronic acid to vitamin C, but that was soon to change. Back in Szeged, Szent-Györgyi was busy attracting staff for his new department. In 1931, he welcomed Hungarian American Joseph Svirbely, who had recently received his PhD degree in chemistry under the direction of Professor Charles G. King at the University of Pittsburgh. King's laboratory was deeply immersed in trying to isolate vitamin C and was finding it to be a difficult problem. Chemists at the giant pharmaceutical company of Eli Lilly had started with vast amounts of dried lemon juice but had failed to isolate vitamin C. Others were also looking to other citrus fruits as a source, but all such juices proved difficult to work with because they also contained many other troublesome components. The difficulties were further compounded by the finding that vitamin C was easily destroyed by heat and often lost its potency with time. There was great scientific attention being paid to the problem of isolating vitamin C in the United States, particularly since many thought that the winner of the race might well receive a Nobel Prize in Physiology or Medicine. This idea was further enhanced by the fact that no American had as yet had ever been so fortunate.

Svirbely, fully acquainted with King's work, suggested to Szent-Györgyi that he test the activity of the remaining few crystals of hexuronic acid that were left over from Szent-Györgyi's previous Mayo Clinic work to see if they had any anti-scurvy activity. Szent-Györgyi supported the suggestion but in a casual, offhanded manner. He hated animal work because "their systems are too complicated." He added that he had never been interested in vitamins or, for that matter, in nutrition generally, dismissing the study of nutrition as "women's work." It seems that his mind remained committed to studying the role of the oxidation-reduction chemistry.

Svirbely developed a sound scientific design for his experiments. He used guinea pigs, and since the vitamin was destroyed by heat, he autoclaved the basic guinea pig feed to destroy whatever vitamin C it might have contained. He then added very small amounts of the hexuronic crystals to the autoclaved ration that was fed to one group, while a second group received only the vitamin C–free diet. Within a month, the animals receiving the hexuronic acid were well and healthy, while the others were dying of scurvy. These critical experiments sealed the link between the adrenal hexuronic acid and vitamin C.

It was an exciting breakthrough. Szent-Györgyi urged Svirbely to write immediately to Professor King telling him of these new findings, since Svirbely had already informed Szent-Györgyi that King was also seeking the link between hexuronic acid and vitamin C. The letter also informed King that a paper by Szent-Györgyi and Svirbely was being submitted to

Nature. King's letter in response stated that he was delaying sending a letter to *Science* announcing the same result, pending the completion of confirmatory results.

On April 1, *Science* published King's letter, announcing that he had discovered that hexuronic acid and vitamin C were identical. His letter did not contain any descriptions of experiments, no supporting data, or any mention of the work of Szent-Györgyi and Svirbely, who had submitted a detailed paper to *Nature*, which was published eighteen days after King's letter. It would take two more months before King published his detailed results. But the battle for priority was on.

King's *Science* letter was widely touted as establishing his priority of discovery, at least in the opinion of American scientific circles. Many thought King would be the first American to win the coveted Nobel Prize in Physiology or Medicine. On April 5, 1932, *The New York Times*, taking up the cry of the Pittsburgh press, published a front-page article with a headline that read "Pittsburgh Professor Isolates Vitamin C."

Even though the *Nature* article had attracted some attention, which the American public largely ignored, they became outraged at the idea that a Hungarian should claim priority! Reading King's *Science* letter, Szent-Györgyi was astonished and infuriated that a scientist of King's stature would publish the Hungarian's great discovery as his own. To him, it was a case of out-and-out plagiarism. Svirbely, seeking to avoid becoming embroiled in a priority controversy, kept the contents of the letter he had received from King strictly to himself, and when he left Szent-Györgyi's laboratory soon after, he took the letter with him. Subsequently he never revealed any specific details of the letter that would have clarified the matter of specific dates, and he maintained his silence on the matter for many years.

During the years that followed, King continued to steadfastly maintain that he had received Svirbely's letter only after he had submitted his letter to *Science*, a position that astonished Szent-Györgyi. Fifty years later, Ralph W. Moss, the author of *Free Radical*, a splendid book about Szent-Györgyi's life, contacted Svirbely, now in retirement, and was able to read the critical letter that clarified the situation.

The letter, dated March 15, 1932, clearly acknowledged King's prior receipt of Svirbely's letter and specifically mentioned that he had not as yet mailed his letter to *Science*. Some think that Svirbely had maintained the letter as secret because, as a young scientist starting out on a career, he wanted to avoid becoming involved in a major scientific dispute that involved a battle between a distinguished American and a Hungarian. He sensed his involvement in such a dispute might seriously harm his future

life and career in the United States, particularly considering the great animosity that many Americans felt towards Szent-Györgyi.

A New Source of Vitamin C

Szent-Györgyi now had a new problem. He was unable to repeat his earlier
work in isolating vitamin C because he had no adequate supply of adrenal
glands available in Szeged. Searching for an alternate source, he tested a
wide variety of plants, but all yielded negative results. A stroke of luck put
him on a new direction, which he related in his own words: "One evening
my wife, Nelly, gave me a dish of peppers for supper, but although I did
not feel like eating it, I had not the courage to tell her so. As I looked at the
peppers, it occurred to me that I had never tested peppers. I told my wife I
would not eat them, and I took them to the laboratory next door." By mid-
night, he had determined that the peppers contained five times the amount
of vitamin C that was present in orange juice, and it was easy to extract.
This tale became known among Szent-Györgyi's friends as "husband's cowardice." Since Szeged was the capital of Hungarian peppers, a large supply
was readily available.

Szent-Györgyi quickly converted his laboratory into a small pilot plant
for the production of vitamin C from peppers. When full production was
reached, one and one-half kilograms of vitamin C were being produced
each week. This was in stark contrast to the thirty milligrams that he had
isolated from the adrenal glands at the Mayo Clinic several years before.
During the next few weeks, the institute changed completely from being
research laboratories to becoming an industrious family enterprise producing paprika juice. Horse-drawn carts arrived with huge heaps of peppers. Everyone, including Szent-Györgyi's wife and daughter, was busy
processing paprika, and townspeople were hired as more hands were
needed. Large fifty-liter balloon flasks were filled with the juice and closed
tightly until the vitamin C could be extracted. A new name was given for
vitamin C: ascorbic acid, the *a* meaning "against" and *scorbic* for the scorbutic (scurvy) disease.

Even though there was a large fortune to be made, Szent-Györgyi
decided not to file for a patent, preferring to give away the vitamin C as
well as the process he had perfected to isolate it. (King had filed for a patent in the US, but ultimately it was rejected.) But Szent-Györgyi had little
success using ascorbic acid as therapy. He treated a Belgian prince who was
suffering from an unknown fever, as well as some other dignitaries with
a variety of other disorders. But his claims of a miracle drug were short-
lived. Szent-Györgyi did obtain a patent on a canned sweetened paprika
as a health food, which he named Pritamin. The royalties from the sale of
Pritamin were used to support the departmental research.

The Nobel Prize

In 1937, Szent-Györgyi was notified that he was the recipient of the Nobel Prize in Physiology or Medicine. The prize committee had argued long and acrimoniously about selecting Szent-Györgyi, so much so that after the final meeting, when the chairman, Hans Christian Jacobaus, came out to make the announcement, he fell dead on the spot with a heart attack. The award carried $40,000 and a gold medal. In Szent-Györgyi's own words: "The Nobel Prize was the only big lump sum of money I have ever seen, I had to do something with it. The easiest way to drop this hot potato was to invest it. Since I knew World War II was coming, I was afraid that if I bought shares that would rise in war, I would wish for the war. So I asked my broker to buy shares that would go down in the event of war. I lost money but I saved my soul."

Time magazine dubbed the award "The Paprika Prize." The Pittsburgh press again took up the cause of King, branding Szent-Györgyi's award as a theft of Dr. King's secrets. Although this dispute eventually cooled down, Szent-Györgyi, the proud Hungarian, felt his honor had been besmirched. Even decades later when the whole matter had faded from the public eye, he was still pained by the affront.

Later Life

Szent-Györgyi was still a young man, just forty-four in 1937, when he received the Nobel Prize. Unlike many other Nobel laureates, he still had a long future ahead of him. Leaving vitamin C research, he launched into biochemical studies of cells and into fundamental studies of how muscle contracts. But war clouds increasingly occupied Szent-Györgyi's mind and his activities. He was a well-known figure in Hungary, and when the war finally came, he became the leader of the anti-Nazi movement in Budapest. He approached the British as the champion of a peace initiative, but it fell on deaf ears. Hitler was enraged when German intelligence learned of it, and he ordered Szent-Györgyi to be captured and brought to Berlin. Szent-Györgyi moved from place to place to avoid being arrested.

The Swedish Embassy in Budapest was instrumental in hiding Szent-Györgyi from the SS troops. When the Russian Army captured Budapest, they took Szent-Györgyi into protective custody. He was quite surprised at this move, since he had opposed the Russian invasion of Finland and had gone so far as to donate his Nobel gold medal to the Finns as a gesture in helping them fight off the Soviet invaders.

After the war, Szent-Györgyi became an important political leader in Hungary, but he grew increasingly unhappy with the Soviet rule, and he finally decided to come to America. At first he was a displaced person, and because of his left-wing leanings, he became a target for Senator Joseph McCarthy's anti-Communist witch hunt.

Szent-Györgyi moved to Woods Hole, Massachusetts, to join the expanding group of scientists working in the Marine Biological Laboratory. Thanks to financial support from friends and later from granting agencies, he established the Institute for Muscle Research in the Marine Biological Laboratories. He later became an American citizen, and dressed as Uncle Sam, he led the Woods Hole parade celebrating the national bicentennial. Szent-Györgyi died in October 1986 at the age of ninety-three. He said that in his hometown, the cheapest funeral you could get consisted of taking a candle in your hand and going out to the churchyard. His own funeral consisting of a graveside ceremony that was simple.

Comment

Szent-Györgyi's discovery of hexuronic acid in 1927 is a classic instance of a searcher finding something, the importance of which he was totally unaware. During his work at the Mayo Clinic, he discovered that adrenal glands were a rich source of hexuronic acid, of which he crystallized about thirty milligrams. He sent half of this small supply to Professor Norman Haworth, a distinguished carbohydrate chemist, for analysis, but it wasn't enough for a comprehensive test. He lost interest in the compound when it failed to be a new adrenal hormone.

Several years later, he came to see its importance, thanks to Joseph Svirbely, a newcomer to the laboratory, who suggested that he test the few milligrams of crystalline hexuronic acid that remained from his Mayo Clinic work for its anti-scorbutic properties. This simple experiment proved the identity of hexuronic acid as vitamin C. Now faced with finding a new source of raw materials for isolating vitamin C, Szent-Györgyi's "husband's cowardice" over refusing to eat peppers for dinner led to his discovery of a large new source. These were the milestones that paved the way for Szent-Györgyi's winning the 1937 Nobel Prize.

REFERENCES

Carpenter, K. J. *The History of Scurvy and Vitamin C.* Cambridge: Cambridge University Press, 1988.

Moss, R. W. W. *Free Radical: Szent-Györgyi and the Battle over Vitamin C.* New York: Paragon, 1987.

Szent-Györgyi, A. "Oxidation, Energy Transfer, and Vitamin." Nobel Lecture, Nobel Prize in Physiology or Medicine, 1937.

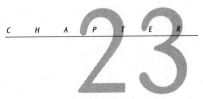

Vitamin D: Bowlegged Lion Cubs

The danger of too little sun is enormous.

—*Dr. Michael R. Eades*

Introduction

I n the fourth century BC, the Greek historian Herodotus observed that the bones of men could differ greatly in their substance. On the battlefield where Cambyses II defeated the Egyptians in 525 BC, the bones of the dead of the two armies lay in separate areas. Herodotus noted that the skulls of the Persians were so fragile that they could be broken if struck by a pebble, but the skulls of the Egyptians were strong and could scarcely be broken when struck by a stone. The Egyptians told Herodotus that their men were always bareheaded beginning in early childhood and were therefore always exposed to the sun. The Persians, on the other hand, covered their heads with turbans, which limited their exposure to sunlight.

In 1650, Francis Glisson, Regius Professor of Physic at Cambridge University, published *Tractatus de rachitide, sive morbo puerili* (A Treatise of the Rickets), in which he described the clinical pictures of the disorder. Glisson stated that the term *rickets* was derived from the Old English word *wrikken*, which meant "to bend or twist." The disease, which he believed to be new, had in his time become common in many of the cities of England. The growth of the industrial cities and towns during the eighteenth and nineteenth centuries proved to be breeding grounds for two major diseases of the poor, tuberculosis and rickets.

Large numbers of families moved from rural areas to the cities for jobs in the dawning Industrial Revolution. They lived in crowded tenements in parts of the cities where exposure to sunlight, especially in the winter, was limited.

The Clinical Picture

Rickets especially affected infants and young children and produced major skeletal deformities. The bones were poorly calcified, and if the child survived, he was left with an odd shaped head and terribly curved deformities of the legs. Standing caused the weight of the body to produce deformities of the weakened bones of the legs, producing bowlegs or knock-knees. The head grew producing a shape that vaguely resembled an egg. Small bumps appeared at the junction of each rib with the breastbone producing the "Rachitic Rosary." (*Rachitic* is the adjective of rickets.) The teeth were poorly calcified. Women who had suffered rickets in childhood were cursed with deformities of the pelvis that created major handicaps for childbirth. In the crowded northern cities of Britain, Europe, and the United States, rickets, often called the "English Disease," was rampant.

In the first decade of the nineteenth century, nearly half of all hospitalized children under three years of age in London, Paris, and New York suffered from rickets. Charles Dickens, always an astute observer, described the late deformities of rickets in the person of Miss Jenny Wren, the "quaint little person of the house" in his novel *Our Mutual Friend*. In Glasgow, Scotland's largest city, rickets was particularly serious. In 1908, Professor Leonard Findlay of the Glasgow University Medical School found that puppies fed on oatmeal and whole milk developed rickets only if kept indoors. If they were walked outside, rickets did not develop. Findlay was convinced that since the disease was most prevalent in the crowded slum areas of the city where young children were mostly kept indoors in unhygienic conditions, it was due to lack of fresh air and exercise. He would continue to support his environmental theory for the next fifteen years, although he was always lacking a strictly critical scientific approach.

A few investigators approached the question from another direction. They picked up on the earlier and nearly forgotten clue about the effectiveness of sunlight. In 1892, British doctor and missionary T. A. Palm observed that there was an inverse correlation between the geographic distribution of rickets in various regions of the world and the amount of sunlight in those regions. In 1919, Dr. Kurt Huldschinsky carried out a singularly creative experiment when he cured rachitic children using artificially produced ultraviolet (UV) light. Two years later, Alfred Hess in New York showed that exposing rachitic children to sunlight could cure them.

Edward Mellanby

Theories as to the cause of rickets varied among overcrowding in tenements, lack of exercise, infection, and food toxicity, but all remained critically untested. Edward Mellanby would be the first to conduct a well-controlled experimental study of the role of diet. In 1824, a French doctor found that cod-liver oil cured rickets, but the more orthodox medical circles in Europe, Britain, and the United States largely ignored his report. As a student at Cambridge in the first decade of the twentieth century, Edward Mellanby (later Sir Edward) came under the influence of the great biochemical nutritionist Frederick Gowland Hopkins (later Sir Gowland; Nobel laureate, 1929), who was deeply involved in research aimed at defining the roles of "accessory food factors" in the diet. Hopkins found that laboratory animals fed sufficient carbohydrates, fats, and protein, all in pure form, grew slowly unless small amounts of other "accessory food factors," later identified as vitamins, were added to the diet.

Following graduation, young Mellanby received a small grant from the newly formed Medical Research Council. Since he had no specific experimental plan in mind, the council suggested that he explore the "defects in the processes by which foods yield energy and the role of these defects as a casual factor of rickets." This stemmed from a theory at the time that diminished oxygen use was an important factor in producing rickets. There was no thought of any accessory food factor being involved, even though Hopkins was a member of the council that approved Mellanby's award.

Mellanby faced the problem of not having a method for producing rickets in young dogs, the only animal then known to develop rickets. Following the council's suggestion, he spent two years trying to manipulate the oxygen use by the animals by administering various drugs, in addition to trying "all kinds of funny things," such as depriving limbs of their blood supply or their lymphatic drainage. All of these attempts failed.

He then recalled that a diet very high in protein caused a 30 percent increase in oxygen use, a well-known effect called specific dynamic action, discovered in 1854. If diminished oxygen use was a casual factor in the development of rickets, Mellanby reasoned that an increase in oxygen consumption should improve the density of bones and hence might be a clue to preventing or healing rickets.

Mellanby was startled at the results of the animals that had been fed large quantities of meat. Contrary to his expectations, the puppies developed rickets, the first animal species to develop the disease under laboratory conditions. It was a critical point in turning his mind from oxidation to food intake. Mellanby decided to explore foodstuffs other than meat to see if any could produce rickets. Among others, he tested oatmeal by feeding puppies a diet existing exclusively of oatmeal porridge. He may have chosen porridge because of some experiments performed by Dr. Leonard Findlay of the Glasgow University Medical School in 1908. Findlay reported that puppies fed on oatmeal and whole milk would develop rickets if kept indoors, but would remain healthy if given exercise by being taken out for walks without any change in their diet. Findlay also pointed out that in Glasgow, where rickets was rampant, the disease was most prevalent in the crowded slum areas of the city where young children were mostly kept indoors in unhygienic conditions, and found no reason to believe that the disease was linked to diet consisting mainly of porridge. For the next fifteen years, Findlay would steadfastly stick to this theory, even as evidence to the contrary was accumulating.

A Lucky Break

During his experiments, Mellanby had kept the puppies indoors instead of having them spend time outside in the dog pen. Mellanby concluded, "It therefore seems probable that the cause of rickets is a diminished intake of an anti-rachitic factor which is either fat-soluble A or a factor which has a similar distribution to fat-soluble A." In retrospect, Mellanby's dogs had been subjected, unintentionally but fortunately, to a routine of not being allowed to spend time outside during the day. Had he followed the customary practices of keeping dogs, which called for allowing the animals daily outdoor time in the sun, he would have missed his crucial demonstration of the need for the anti-rachitic factor. Although Mellanby's work clearly established the role of diet in the cause of rickets, it left open the question of the nature of the "anti-rachitic" factor, whether it was vitamin A or some other closely related substance. That question was later resolved by Professor Elmer McCollum, a pioneer biochemical nutritionist who showed that when cod-liver oil was treated to destroy its vitamin A content, it still retained its anti-rachitic, fat-soluble factor, which he named vitamin D, the fourth vitamin after A, B, and C that had been discovered.

By chance, Mellanby happened to hear of the earlier experience of the well-known surgeon Dr. John Bland-Sutton (later Sir John), who in addition to his fame as a surgeon had had a lively interest in the habits and diseases of animals. In his student days, Bland-Sutton had been a prosecutor, performing necropsies on animals who had died in the Zoological Gardens, an experience that provided him with a broad knowledge of animal pathology. In 1889, he published a paper reporting that rickets occurred in carnivores, ruminants, rodents, marsupials, and birds. But the disease was most marked in lion cubs, and prior to Bland-Sutton's interest, consecutive litters of lion cubs had developed fatal deformities.

Bland-Sutton instituted a new diet of lean goat meat, crushed goat bones, and milk for both dams and cubs along with cod-liver oil. On this regimen, the cubs grew normally and showed little if any traces of their earlier deformities.

These results were obtained four decades prior to the general acceptance of the use of cod-liver oil in infants and children. Unfortunately, Bland-Sutton never published the details of the diet that he had designed. Although these classical feeding experiments in lion cubs were quoted in many textbooks at the beginning of the twentieth century, they still attracted little attention. A more recent analysis of Bland-Sutton's work has amplified the success of cod-liver oil by pointing out that the bile acid, necessary for the absorption of fat-soluble vitamins A and D in feline species, requires an exogenous intake of taurine. Cod-liver oil supplies both vitamin D and taurine, which allows the absorption of the vitamin D.

Cod-liver oil had been used in children since at least 1824, but medical opinion was never fully convinced of its value. Professor John Howland, a great American pediatrician, was aware that cod-liver oil had been used therapeutically in the past, and yet he withheld his unequivocal approval. The widespread delay in using cod-liver oil for rickets was due to several factors. Some of the reluctance stemmed from the belief at the time that using any animal organs in medicine was frowned upon.

Often, doctors and the public had dismissed cod-liver oil as just another kind of oil, and a rather nauseating one at that. The results of administering cod-liver oil to a child with rickets were not readily evident, perhaps due to the variable composition of the oils then in use as well as the inability to define the proper dose and duration of treatment. But most of all, it was the then current medical paradigm that cod-liver oil was a medicine and not an accessory food factor. A completely new paradigm would be needed for physicians and the public to realize that cod-liver oil contained an essential dietary component. Mellanby would convincingly establish this new mindset.

In 1919, Mellanby, having produced rickets in dogs by feeding them oatmeal porridge, reported that he could cure the rickets in his animals by adding cod-liver oil to the diet. He concluded that rickets is a condition primarily due to the lack of an accessory factor in the diet that is present in cod-liver oil. Using his rachitic puppies, Mellanby evaluated other fat-containing foodstuffs for their curative effect, using cod-liver oil as the standard. Butter proved somewhat less effective, but no vegetable oil had any effect. Mellanby strongly warned against using vegetable oils as substitutes for cod-liver oil. His criticism was specifically directed against a proprietary product then in wide use called "Marylebone Cream," in which the active ingredient was linseed oil.

Mellanby had established the existence of an essential dietary factor, but its chemical structure was unknown, although it had some association with certain fats in the diet. For a time, Mellanby thought that vitamin A, a fat-soluble compound, might be the anti-rachitic factor. But that was found to be in error in 1922 when McCollum showed that the anti-rachitic factor was not vitamin A. McCollum named the new factor vitamin D.

Harriette Chick: The Vienna Study

Immediately following the end of World War I, Europe was plagued with widespread malnutrition as the result of the Allied blockade. Vienna was particularly hard hit, since its only food supply had come from neighboring Hungary, and at the end of the war the newly independent Hungary ceased to supply food across the border to Austria. In 1918, the British Medical Research Committee established a mission to Vienna, headed by Harriette Chick (later Dame Harriette), to study malnutrition in infants and children, and specifically to test the applicability of Mellanby's experiments. Professor Clemens von Pirquet, director of the University of Vienna's Kinderklinic (Children's Hospital), invited the British team to use the hospital facilities, which included a ward of twenty cots for infants and forty more in the Amerikanische Kinderheilstatte, a children's sanatorium also under von Pirquet's direction.

The principal aim of the mission was to ascertain whether the results in Mellanby's puppies held up in infants with rickets under controlled conditions. In all, sixty-eight infants under the age of five months completed the study. Half of the infants received diet I, consisting of milk with added sucrose, while the other half received diet II, which was based on imported full-cream, dried milk, with no added sugar but with the added supplement of cod-liver oil. Serial X-rays of the bones of the arms and legs were used to assess the development of rickets and to ascertain the degree of mineralization of the bones.

All of the infants receiving diet I developed rickets, whereas none of the diet II infants receiving the cod-liver oil developed the disease. It was noteworthy that many of the rachitic babies had developed rickets during their stay in the extremely hygienic conditions of the hospital ward. This observation marked the downfall of Dr. Leonard Findlay's fifteen years of ardent pronouncements that rickets was due to infants being exposed to "slum" conditions.

But there was a surprise in store. As part of its routine practice, the rachitic infants and the control infants were placed on an adjacent outdoor terrace in broad daylight. With the beginning of spring, those babies fed diet I showed rapid healing of their rickets. The UV rays of sunlight could cure rickets! The role of sunlight had been suggested before, but the supporting evidence was often anecdotal, uncontrolled, or based on demographic inferences. None offered the clear proof from studies conducted under the controlled conditions that Chick followed. The role of sunlight explained the well-known fact that babies raised in the crowded, dark inner cities developed rickets in the winter months but often improved in the spring and summer. It wasn't the unhygienic living conditions, infection, lack of exercise, or any food toxicity that caused rickets in these babies. It was the lack of sunlight plus an insufficient intake of vitamin D in the mothers' breast milk.

Chick's study unequivocally confirmed Mellanby's results that a dietary factor was definitely involved. But the results on sunlight were a major breakthrough and were confirmed independently by others in animals and infants.

Reflecting on her work, Chick made the prophetic suggestion that the anti-rachitic factor was synthesized in the skin under the influence of radiation from the sun. It was the first time that the skin had ever been thought of as an active organ that reacts to particular light waves to produce vitamin D, and it was much more than a passive protective covering of the body. This was an entirely new concept at that time, but it would be validated in the next few years of research.

The Mystery of Sunlight

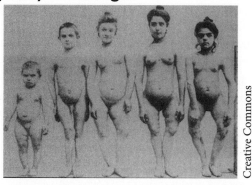

Attention now shifted away from the diet and toward studying the details of the effects of ultraviolet light. No one understood how ultraviolet light worked. The idea that light could cause a chemical to be made by the body was completely foreign. Up to this point, the thinking had concentrated on a fat-soluble component of cod-liver oil that was distinct from the fat-soluble factor vitamin A. But now it appeared an entirely different process also could cure rickets by exposure to sunlight or to irradiation by UV light that simulated sunlight. Was sunlight simply a part of general improvement in general health, or was there something more specific in its curative powers?

Three different experimental results, all within the short time of one year—1924—began to clarify the mystery between vitamin D and sunlight. First, rats suffering from rickets benefited from irradiation by UV light, not only by irradiation of the rats themselves, but also by irradiation of the "air" in the glass jars from which they had been removed and then returned. This mystery was solved when it was discovered that the rats had eaten the irradiated sawdust, feces, and spilled food. Second, the livers of irradiated rats were curative when fed to rachitic rats. And furthermore, liver and muscle tissue from non-irradiated rats could be activated by UV radiation, after which they could promote growth and bone calcification when fed to non-irradiated rachitic rats.

Third, it was well known at that time that certain animal fats such as butterfat or cod-liver oil were anti-rachitic without irradiation. But UV light made foods such as wheat, lettuce, or cottonseed oil anti-rachitic. Irradiated whole milk became anti-rachitic. This observation led to major practical advances in public health, when irradiated milk led to a rapid decline in the prevalence of rickets in infants and children.

What Is Vitamin D?

The search was now on to identify the chemical nature of the substance responsible for these effects. At first, cholesterol was thought to be the most likely, particularly when Alfred Hess found that cholesterol in the skin activated by UV irradiation becomes anti-rachitic. But almost immediately, there was concern as to the purity of the "pure" cholesterol, since a chemical analysis had shown the presence of a minute amount of an "impurity," which was thought to be the source of the anti-rachitic activity. It seemed that it was the immediate precursor of vitamin D, but its composition was unknown.

It was at this stage that Hess in New York invited the famous steroid chemist Adolf Windaus, in Göttingen, Germany, and O. Rosenheim in London, to collaborate with him on the defining the chemical structure of the anti-rachitic product. Windaus and Hess produced a highly concentrated active fraction of the provitamin D from so-called "pure" cholesterol, and determined that it was ergosterol, a steroid widely distributed in plants but not in animals. Simultaneously, Rosenheim found that ergosterol was the provitamin D and was convertible to an anti-rachitic substance by UV light. Irradiated ergosterol was found to be highly active in curing rats with rickets. The irradiated product of ergosterol was purified and crystallized simultaneously by the London team and by the German team, and was named vitamin D$_2$, or calciferol. It showed enormous potency; as little as 0.01 micrograms per day administered to rachitic rats for two months completely cured rickets.

But another problem remained. Ergosterol was widely found in plants, but not in animals. Since animals do not make ergosterol, how did they obtain vitamin D by sunlight? This important problem was not solved until 1937, when Windaus identified a compound, 7-dehydrocholesterol, from the skin of animals and humans and in food sources (such as whole milk or liver) that was convertible to an anti-rachitic substance by irradiation. This irradiated product was named vitamin D$_3$, or cholecalciferol.

A Note on Terminology

Customarily, we define a vitamin as a substance that cannot be made in the body and must be present in small amounts in a healthy diet that is essential for the health of the individual. The dietary source is important, since the body cannot make the vitamin in question. But with vitamin D, there is a different story. The body makes vitamin D in the skin, the largest organ of the body, through the irradiation by the sun's UVB rays (the so-called tanning rays). Once created, it circulates in the blood and in other body fluids as it is delivered to various organs. These attributes fit the definition of a hormone and not of a vitamin. In fact, properly speaking, the active substance is a steroid hormone called cholecalciferol. This terminological nicety is rarely encountered except in the research literature of the field. Vitamin D continues to be the universally familiar term.

REFERENCES

Carpenter, K. J. "Harriette Chick and the Problem of Rickets." *J. Nutr.* 138 (2008), 827–832.

Chick, H. "Study of Rickets in Vienna 1919–1922." *Med. Hist.* 20 (1976), 41–51.

Conlan, R. et al. "Unraveling the Enigma of Vitamin D," in *Beyond Discovery*. http://www.beyonddiscovery.org.

Mellanby, E. "An Experimental Investigation on Rickets." *Lancet* (1919), 107–112.

Part VII
Other Medical Discoveries

24

Insulin: The Role of Insomnia

Diabetes is no longer a death sentence.

—*Frederick Banting, 1923*

Introduction

L eonard Thompson, a fourteen-year-old-boy, was admitted to the Toronto General Hospital in December 1921. Leonard had had severe uncontrolled diabetes for three years, and it had taken a dreadful toll on the boy. He was pale and emaciated, and he weighed only sixty-five pounds. His muscles were flabby, and his dry hair was falling out. His breath smelled of acetone, characteristic of uncontrolled diabetes. The sugar in his blood was sky-high, and his urine was loaded with sugar. The doctors and nurses who attended Leonard felt he was at death's door. They expected him to slip into a diabetic coma and to die within a few days. There was nothing they could do to stop this seemingly inexorable course of events.

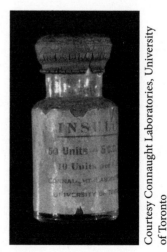

Courtesy Connaught Laboratories, University of Toronto

The First Tests

Across the street from the hospital in the department of physiology of the medical school, two researchers had been hard at work during the previous year in isolating and identifying the pancreatic hormone that controls sugar and fat metabolism. As the results of some highly encouraging experiments in diabetic dogs, the investigators had been able to isolate the cells of the pancreas that produced insulin and to partly purify it.

It was time to try their preparation, called insulin, on a human subject. Leonard Thompson was an ideal candidate. A small dose of a crude extract made from the insulin-producing cells of the pancreas was injected under the boy's skin. Within a few hours, his blood sugar had dropped by less than 20 percent, but the sugar and acetone in his urine remained high. It was a disappointing first test result. Reasoning that the extract was very crude and lacking in potency, a much more purified extract was given twelve days later, this time with dramatic falls in blood and urine sugar and a disappearance of acetone from his breath and his urine. The insulin dose was increased at regular intervals until at the end of two weeks; the boy was much more alert and more active. He looked better and he said that he felt stronger.

As the treatment continued, Leonard developed the appearance and activity of a normal healthy boy, who differed from the others only because of his dependence on insulin. Leonard lived to be twenty-seven years old. This was the first time that severe diabetes had been reversed. It was a miracle! Frederick Banting, a physician, and Charles Best, a medical student, had created this miracle. Countless other diabetics, both children and adults, have followed Leonard Thompson with equally revolutionary results.

Oscar Minkowski

The first hint of a relationship of the pancreas to diabetes occurred in 1869, when Paul Langerhans, a German medical student, looked through his microscope and observed two different types of cells in the pancreas. One of these, the cell that produces the digestive enzymes, was widespread throughout the organ. But the other cell type occurred in clusters that resembled islands 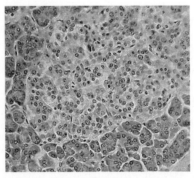 in a sea of the first type of cells. Their function was unknown. There was some speculation that these islets of Langerhans, as they came to be known, produced an anti-diabetic factor that was involved in sugar metabolism, but that idea was far from proven. Even the relationship of the pancreas to diabetes was still in doubt.

Public domain

Oscar Minkowski

In 1889, Dr. Oscar Minkowski of the University of Strasbourg made an historic discovery. Minkowski, working in the medical clinic, needed a journal from the chemistry department library. While he was there, he stopped to chat with his friend Joseph von Mering. Von Mering told Minkowski that he had prepared oil that he believed would treat rickets more effectively than cod-liver oil, because he thought that unlike cod-liver oil, this new oil would not require a pancreatic enzyme for its absorption into the blood.

But there was a big problem with von Mering's idea, since the only way to test it would be to administer it to an animal after its pancreas had been entirely removed. But at the time, no one thought that surgical removal of the pancreas was technically possible. Minkowski, with a touch of youthful bravado, scoffed at this obstacle, and von Mering challenged him to perform the operation. He even offered a dog to be the experimental subject.

The next day Minkowski, with von Mering's assistance, surgically removed the entire pancreas of the animal. His success was not due to being a great surgeon so much as the fact that this particular dog had a highly unusual anatomical location of its pancreas, which luckily allowed Minkowski to remove it completely. The story might very well have ended here if a dog with normal pancreatic anatomy had been used.

After the operation, the dog was tied up in one part of the laboratory, where it soon began to drink large amounts of water and to pass copious amounts of urine. Minkowski scolded his lab assistant for not letting the dog out more frequently in the courtyard to urinate. The assistant replied that this dog often urinated outside and again almost as soon as it got back inside. He added that flies were swarming over the puddles of urine. Minkowski had a suspicion that lacking its pancreas, the dog had developed diabetes with sugar in the urine.

He rushed into the courtyard, dipped the tip of his finger in a puddle, and tasted it. It was very sweet! Back in the lab, he analyzed a sample and found that it contained 12 percent sugar. Minkowski had conclusively confirmed the link between the pancreas and diabetes. He published his results in what is now a classic paper, entitled "Diabetes Mellitus nach [after] Pancreas Extirpation."

Minkowski surmised that the pancreatic tissue he had removed must contain some kind of an anti-diabetic factor. If so, an extract of the removed tissue should reduce the level of sugar in the urine. That experiment didn't succeed, even though he was on the right track. As we will see a little later, the isolation of the anti-diabetic factor would require more sophisticated methods than were available to Minkowski. Nevertheless, the very idea of the existence of such a factor stimulated many other scientists to search for it over the next thirty years. They all failed.

Frederick Banting

Captain Frederick Banting, a Canadian Army orthopedic surgeon return-ing from the war, would achieve the success that had been denied to all pre-vious experimentalists. In July 1920, Banting settled in London, Ontario, and set up a small practice. The practice attracted a pathetically small num-ber of patients and generated such a meager income that Banting took a part-time teaching position in the department of physiology of the London Medical College of the University of Western Ontario.

Banting's assignment was to give a set of lectures on the structure and function of the pancreas. Since Banting was a surgeon and not a physiolo-gist, he devoted many hours in the library, poring over books and papers that dealt with the pancreas in preparation for his lectures to students. In his reading, Banting ran across references to the enigmatic relationship of the pancreas to diabetes, and the speculation that the islets of Langerhans were somehow involved.

Banting was an insomniac. On October 30, 1920, at two o'clock in the morning, unable to sleep, he started browsing through some journals, scan-ning articles that were clinical in nature and far removed from his assigned topic of teaching the physiology of the normal pancreas. Glancing through November issue of *Surgery, Gynecology and Obstetrics*, Banting saw an article entitled "The Relation of the Islets of Langerhans to Diabetes with Special Reference to Cases of Pancreatic Lithiasis" that was written by Dr. Moses Barron, a distinguished pathologist. In his work, Barron had come across several cases in which pancreatic lithiasis, or stone, had completely obstructed the main pancreatic duct, which had led to a destruction of the acinar cells (i.e., the cells responsible for producing the digestive enzymes) but left the islet cells intact.

Banting immediately realized that tying off the pancreatic duct, waiting for auto-digestion to occur, and then extracting the intact islet tissue might be a promising way to isolate the anti-diabetic factor. He was so excited by this idea that he gave up trying to sleep and made a note in his note-book: "Tie off pancreas ducts of dogs. Wait six or eight weeks. Remove the remaining tissue and extract it."

Banting was so taken with the idea that he decided to abandon his med-ical practice and to pursue it in the laboratory. But to do so, he would need laboratory facilities, and the medical college had none available for his use.

Perhaps his old medical school at the University of Toronto might be more favorably inclined to help. In November 1920, Banting set out on the hundred-mile trip from London, Ontario, to Toronto in his creaky old Ford car in order to call on the esteemed Dr. John James Rickard Macleod, professor and chairman of department of physiology.

The interview was brief and uncomfortably tense. Banting presented his ideas, but Macleod, fully aware of the many previous unsuccessful attempts (including some of his own) to find the pancreatic factor, scornfully dismissed Banting. He told Banting that his idea would surely fail. After all, Banting, being a mere orthopedic surgeon, had no credentials or experience as a physiologist. Not to be outdone, Banting made a second trip during the Christmas holidays

Dr. Macleod

Public domain

and again, the strong-minded Macleod summarily refused his request. Banting was furious at being treated in such a hostile and demeaning manner, particularly since he was a graduate of the university's medical school. It was the beginning of an animosity between the two men that would grow increasingly bitter with time.

Charles Best Joins Banting

On Banting's third trip to Toronto, Macleod, having been urged in the meantime by some of Banting's old medical school friends, finally consented to let him use his own laboratory while he was away on a long summer fishing holiday in Scotland. But the offer didn't include any funds for animals or for laboratory supplies. Macleod thought that the summer's laboratory experience would teach Banting the folly of his idea.

Since Banting had no training in chemistry, Macleod suggested the names of two medical students who could fill that gap. The two students were seniors in the physiology and biochemistry honors program and had been working as research assistants, during which time they had learned the laboratory tests needed to determine sugar in the urine and blood. The two tossed a coin to decide who would be the first to take his summer holiday, and upon his return, he would take over as the assistant to Banting. Charles Best lost the coin toss. By the middle of the summer, Best was so enthralled with the work that he opted to forgo his holiday entirely.

Banting and Best began their work on May 17, 1921. Macleod gave suggestions about the chemical treatment of the harvested islet cells. They had been working for about a month, consulting intermittently with Macleod about problems that arose. Macleod left in mid-June for his three-month holiday in Scotland. The only financial support for the work was Banting's meager savings from the sale of his house and furnishings and his old Ford car. Every cent went toward the work. Best worked with no pay, and to save money, the two scouted poor neighborhoods, buying dogs from owners who needed the money.

After several failed attempts over the miserably hot summer of 1921, Banting succeeded in tying off the duct of a single dog. He and Best waited several worrisome weeks for the destruction of the pancreatic tissue to be complete. Banting then removed the shriveled pancreas and prepared an extract in the hope that it contained isletin, the name they had given to the anti-diabetic factor, which they derived from the islets. Banting successfully removed the pancreas of a second dog that would be the test animal to see if extracts of the islet cells lowered the blood sugar. After an injection of isletin extract into a vein, the dog rapidly regained consciousness, and its blood sugar level fell to normal. They had found their isletin! The lively dog was given free run in the laboratory for the next week, during which it received all of the remaining extract before succumbing to a fatal diabetic coma.

When Macleod returned, he could hardly believe that these two laboratory neophytes had been successful in confirming Banting's original idea. But he was skeptical of Banting's laboratory notes, and he not so silently impugned Banting's personal integrity. A rancorous argument erupted. Tempers cooled, and Banting and Best were able to continue their work in the hostile environment of Macleod's laboratory. Macleod required Banting to perform more experiments as well as to work on projects unrelated to "isletin," an assignment that further increased his animosity toward the arrogant professor.

J. J. R. Macleod and J. B. Collip

As chairman, Macleod controlled all of the resources of the department and thus was able to absolutely dominate Banting, whom he regarded as an "impecunious, summer laboratory worker." But as the work progressed, and new results matched the earlier successes, Macleod was finally persuaded of the validity of the work, and he put his full support behind extending the studies. He recruited J. B. Collip, a gifted biochemist, who prepared extracts that were administered to a series of diabetic dogs. A certain dose of extract restored the blood sugar to normal, but too much resulted in listlessness, and in extreme cases, the dog developed convulsions. Feeding or injecting sugar rapidly overcame these effects. Collip used his chemical wizardry to obtain a more highly purified extract, which he christened *insulin* (from the Latin *insula*, island), acknowledging its source as the islets of Langerhans.

But Collip declined to reveal any of the details of his chemical procedures to either Banting or Best. He abruptly announced that he and he alone would go ahead and patent his secret procedures. Hearing this, the enraged Banting grabbed Collip by the coat, slammed him down on a chair, and came close to "knocking the hell out of him." Shortly thereafter, the four principals, Collip, Banting, Best, and Macleod, agreed that no one would seek any individual patents. With Macleod's support, Collip made good progress in techniques for purifying insulin, but it was clear that his laboratory could never meet the needs of the many thousands of diabetics whose lives depended on it.

The word of this new discovery as a miracle drug for diabetic children led to an avalanche of letters from parents of diabetic children. They had learned that this new medicine could restore the lost weight of their affected children by restoring their bodies to normally use sugar. One such letter came from Mrs. Charles Evans Hughes, wife of the famous jurist Charles Evans Hughes, who wrote, "My dear Dr. Banting: I am very anxious to know more of your discovery." She went on to describe her daughter's case: "She is pitifully depleted and weighs 45 pounds." Banting was unimpressed with all such messages, since the supplies of insulin at that stage were scarcely able to meet the needs of the patients being treated in the hospital. But soon thereafter, he relented and offered insulin for Elizabeth Hughes. Evidently, Justice Hughes had pulled some high-level strings that were sufficient to change Banting's mind. Elizabeth became the poster child of insulin. She regained all of her lost weight and had a normal life. She married and had several children. Few if any of her friends knew that she was a diabetic. She died at the age of seventy-three. During her life, she had received an estimated 42,000 insulin shots.

Macleod versus Banting

In November 1921, Macleod asked Banting and Best to present their research to students and staff at the Physiological Journal Club. Best was to show charts of the dogs, and Banting was to describe the work. In introducing the pair, Macleod preempted everything that Banting and Best were prepared to say. To make matters worse, Banting was far from being a polished speaker, and his garbled presentation went poorly. He was irritated that Macleod kept referring in his remarks as "we," which left the audience with the strong impression that Macleod, not Banting and Best, had made the discovery.

Macleod had submitted an abstract for oral presentation to an important national meeting of the American Physiological Society to be held in
late December 1921 at Yale. Macleod had attached the names of Banting
and Best "by invitation," which was customary way to list nonmembers of
the society. Banting again stumbled as he presented the results, and again
Macleod intervened with numerous clarifying remarks about the work
that "we" had done. Banting was furious at Macleod's remarks, and his
past enmity toward Macleod was further inflamed when he realized that
Macleod was angling to have the medical community, and eventually the
lay public, recognize him as the true discoverer.

In his defense, Macleod must have felt that as chairman of the department and titular head of the laboratory, he was the brains behind the
experiments of Banting and Best. He began to cultivate press coverage for
"his" discovery. Banting hated the press and was incensed by some unfair
newspaper articles that gave virtually all of the credit to Macleod, including
even the idea of tying off the pancreatic duct! A livid Banting confronted
Macleod and asked him to set the record straight in writing. Macleod
penned a letter to the press hinting that Banting's role was merely confined
to simply tying off the duct. His letter went on to imply that Banting's inex-
perience in physiological research could hardly have ensured success in
discovering insulin.

Among the attendees at the Yale meeting was Dr. George H. A. Clowes,
the recently appointed director of research of the Eli Lilly Pharmaceuticals
Company. Clowes immediately recognized the enormous sales potential
of insulin, and he negotiated agreements between Lilly and the University
of Toronto in which Lilly would have rights to manufacture insulin and
the university would receive royalties. Within a year, Lilly was manufacturing insulin in sufficient quantities to meet the American demand, and
many companies in other countries followed suit.

Banting and Best had begun their experiments in May; and by November
they had finished their first paper for publication and submitted it under the
title "The Internal Secretion of the Pancreas" to the *Journal of Laboratory
and Clinical Medicine* for its February 1922 issue. The paper contained the
following landmark observations: "In the course of our experiments, we
have administered over seventy-five doses of extract to ten different diabetic animals. Since the extract has always produced a reduction in the
percentage sugar in the blood and of the sugar excreted in the urine, we
feel justified in stating that the extract contains the internal secretion of
the pancreas." These words ushered in the revolution that saved Leonard
Thompson's life, which would eventually follow for the benefit of countless
numbers of diabetics.

More angry quarrels ensued between the two strong-minded men about how much credit should be given to Banting and Best. Macleod's success with the press and Banting's utter disdain for it finally led to irreconcilable differences between the two. Part of the problem was the totally different styles of the two men. Banting was from the "rough" country and spoke "haltingly," while Macleod had "upper class" polish and spoke "beautifully."

The Nobel Prize

In late 1923, the Nobel Prize Nominating Committee had begun deliberations for the selection of the 1924 Nobel Prize. August Krogh, a famous Danish scientist and Nobel laureate, visited Macleod's laboratories and was impressed with what he saw. In fact, the relations between these two were warm to the point that Krogh was a houseguest of the Macleods during his two-day stay. In his report, Krogh proposed that the prize be divided equally between Macleod and Banting. The Nobel Committee agreed. Although the Nobel statutes allowed for an award to be shared three ways, Charles Best, whom the committee seemed to feel was just a mere a medical student, was ignored, despite the opinion of many, including Banting, that he deserved to be recognized as an equal in the discovery.

Banting exploded when he heard that Macleod was a co-recipient of the prize. Colleagues recalled that they had never seen him so angry. He stormed up the steps of the Physiology Building, determined to do Macleod personal harm, but cooler heads eventually prevailed. Banting promptly dispatched a telegram to Best, who was giving a lecture at Harvard, informing him that while he, Banting, had won the Nobel award along with Macleod, he would divide his prize money equally with Best. In his anger, Banting refused to travel to Stockholm to collect his share of the prize money.

Not to be outdone, Macleod announced a few days later that he would share his prize money "equally" with Collip. According to some accounts, Macleod later said that Collip would receive his "fair share," and that his earlier announcement had been a "misstatement." In 1981, Professor Rolf Loft, a former chairman of the Nobel Committee for the Prize in Physiology or Medicine, told the world that in his opinion, the award to Banting and Macleod that neglected the work of Best was the worst error of ommission that the committee had ever made. He dismissed Macleod as a "manager and promoter who put Collip and the Lilly Company into business."

In 1928, Macleod resigned his post as professor of physiology, to be replaced by the twenty-nine-year-old Charles Best. Macleod had happily accepted the appointment of regius professor at the University of Aberdeen. He left Canada an embittered man, having endured years of ceaseless arguments with Banting. As he was walking to the train that would take him away from Toronto, a friend asked him why he was walking with a shuffling gait. He answered, "I'm wiping away the dirt of this city." Once in Aberdeen, he curtly dismissed all inquiries about the "trouble in Toronto" without further comment. Macleod died in Aberdeen at the age of fifty-nine.

Banting richly deserved the many honors he received, including a knighthood. His name still graces the Banting Foundation, which provides grants to beginning researchers; the Banting and Best Department of Medical Research; and the Banting Institute at the University of Toronto. Banting became and still is one of the most honored persons in Canadian history. In 1939, Banting reenlisted in the Canadian Army and was soon deeply involved in the development of pressure suits for pilots. On a trip to England to demonstrate the suits, his plane crashed, and Sir Frederick was killed. He was fifty years old. Charles Best became a great physiologist and made many subsequent contributions; he was nominated several times for a Nobel Prize for work he had done after his partnership with Banting. The Charles Best Department of Physiology at the University of Toronto was renamed in his honor.

Comment

Chance played a large role in the discovery of insulin. Minkowski ambled over to the library of the chemistry department, and while there, he had a chance conversation with von Mering. Responding to von Mering's challenge to perform the removal of a pancreas on a dog, Minkowski luckily operated on a dog whose rare anatomical location of its pancreas was ideal for an easy total removal by an amateur surgeon. Minkowski knew that the high sugar content of the dog's urine was the direct result of the loss of its pancreas, but he was unable to identify the mysterious factor responsible.

Had Banting's orthopedic practice succeeded, he would never have needed the part-time position in the medical school, teaching the physiology of the pancreas. We can safely assume that Banting, acutely aware of his precarious financial circumstances, would strive to become the best teacher in the department by devoting his energy and time to learning as much as he could about the physiology of the pancreas. Otherwise he would have never seen and understood the significance of Barron's paper dealing with obstruction of the pancreatic ducts that was so crucial to the experiments that he and Best would carry out. It was Banting's alert mind, occupied with learning about the pancreas, that connected Barron's observations to a plan for experiments that would unlock the secrets of the islet cell production of insulin.

Banting and Best had no idea how fortunate they were that insulin from the dog was effective in humans. In 1955, Professor Frederick Sanger identified 51 amino acids in the insulin molecule that are arranged in a unique order, much as the letters in a word are arranged in a specific order. Sanger received the Nobel Prize for this work. When human insulin was compared to dog insulin, the differences were minute. Had they been larger, dog insulin might have been completely ineffective in human diabetics. Had that been the case, Banting and Best might well have abandoned their search altogether during those heady months of the summer and fall of 1921.

REFERENCES

Bliss, M. *The Discovery of Insulin.* Chicago: University of Chicago Press, 2007.

Cooper, T., and A. Ainsberg. *Breakthrough: Elizabeth Hughes, the Discovery of Insulin, and the Making of a Medical Miracle.* New York: St. Martin's Press, 2010.

Zucker, A. "Rediscovery of the First Miracle Drug." *The New York Times*, October 5, 2010.

C   H   A   P   T   E   R

25

Viagra: A Pleasant Surprise

Blind discovery is a necessary condition for a scientific revolution.
— Aharon Kantovich and Yuval Ne'eman

The Beginnings

The research teams based in Pfizer Laboratories in Sandwich, England, had been studying the chemistry of a naturally occurring substance called Atrial Natriuretic Factor (ANF). This substance was especially interesting because it produced an accelerated excretion of sodium in the urine, and hence might be applicable to various disorders characterized by sodium retention—for example, heart failure, where sodium retention is basically the cause of the edema.

From a biochemical viewpoint, ANF was known to stimulate production of a chemical named GMP in the tissues, but it was rapidly degraded by another chemical, abbreviated PDE. The researchers reasoned that if a block for PDE could be found, the GMP would rise, which in turn would enhance the renal excretion of sodium as well as dilate blood vessels. At the time, 1985, the chemistry of PDE was known. (Later this PDE was renamed PDE-5 as the fifth of the eleven known PDEs, each of which is specifically numbered.) The chemistry team was able to create a potent inhibitor, and this formed the basis of what was tagged as UK 92480, which would be eventually be named Viagra.

But first they had to design studies that tested the effectiveness as well as the safety of this new drug. They started with healthy volunteers. Several men may have experienced penile erections, but at the time, no one thought that that was noteworthy. In 1992, W. Michael Allen, a project manager at Pfizer, answered a phone call from a doctor in the small Welsh town of Merthyr Tydfil, where a trial of UK 92480 in healthy men was underway to assess any side effects. The doctor reported some side effects of indigestion, back pain, and others, but he added that several men reported having spontaneous erections. Another trial aimed at the effects of UK 92480 on men who received nitrate-containing drugs such as nitroglycerin, which is the classical therapy for angina. Studies on men taking such a drug when given UK 92480 indicated that the new drug showed an unexpected and significant fall in blood pressure. This was very discouraging, and the quest for an angina drug would be discontinued in favor of a serious study of the erections. (A panel of experts in heart disease have published recommendations for the safe use of Viagra in cardiac patients.)

There was already anecdotal evidence that UK 92480 was involved in producing erections of the penis, but carefully designed trials in men were required. The rules of any trial of a drug call for a statistical comparison of the expected effects with the effects of the placebo. In addition, the range of dose versus effect should be determined. The investigator must monitor side effects in relation to dose and duration of the test period. The Pfizer scientists were well aware of these rules and proceeded accordingly.

One of the first trials of UK 92480 was in men with erectile dysfunction (ED) while watching erotic movies; it was very encouraging. Healthy volunteers who received multiple doses for ten days had the same "side effects" as the placebo group. Some of the volunteers experienced headache, muscle aches, and indigestion, perhaps related to the dose or the duration of the trial. None of these was deemed to be serious enough to stop further work on the drug. At this point, Pfizer saw the potential for a new blockbuster drug, and it devoted massive monetary resources to make the drug a reality.

More than three hundred patients from the UK, Sweden, and France were treated for a four-week period while receiving three different doses and a placebo. When all the data collection was complete and analyzed, Ian Osterloh, one of the principal investigators, exclaimed, "The results exceeded our wildest dreams." There was a clear effect of dose, with 90 percent responding at the highest dose. The drug was well tolerated, and there were only a few dropouts.

Mechanism of Action

Many chemical systems in the body are activated by chemical messengers that are produced in one type of cell and then enter different neighboring cells, where they may produce another mes-

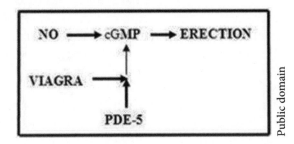

senger, which speeds the final action of still another type of cell to produce a result. In the case of a normal penile erection, the first messenger is nitric oxide (NO) that is made and released by nerve endings. NO produces a second messenger, cGMP, which is regulated by PDE-5.

PDE-5 is present in erectile tissue. UK 92480, soon to be named Viagra, binds itself to PDE-5 in that site and inactivates it, thus enhancing cGMP to keep the blood vessels of the penis open until they are completely filled. The erectile tissue of the penis, the blood vessels of the lungs, and the blood platelets are the only tissues in the body that contain PDE-5. The retina of the eye contains PDE-6, and a slight cross-reaction with PDE-5 may account for some of the visual disturbances of patients taking large doses of Viagra.

During the late 1980s and early 1990s, there was a growing awareness of the important role of NO in mediating various biological systems. NO is not to be confused with N_2O, nitrous oxide, the anesthetic gas. In 1998, the Nobel Prize was awarded to Robert F. Furchgott, Louis J. Ignarro, and Ferid Murad for their work in discovering the importance of NO, which is a short-lived, endogenously produced gas that acts as a signaling molecule in the body.

The role of NO fitted nicely into the cGMP-PDE system, since it is produced by nerve endings and diffuses into the cells lining the erectile tissue of the penis, stimulating them to produce cGMP. The role of nerves in this process may explain why sexual stimulation starting with the brain is required for setting off the entire process.

Viagra Gets Its Name

Long before the final approval of Viagra, the search was on for a suitable trade name. Trying to name a trademarked drug is rarely a sudden brilliant stroke of luck. Rather the trade name must successfully pass a number of hurdles before receiving final approval by the US Food and Drug Administration as well as by the US Patent and Trademark Office.

Every drug has three names: a chemical name based on the compound's chemical structure; a generic name, which is the drug's official name throughout its lifetime; and a trade or proprietary name that is used by the pharmaceutical company to identify the drug to the public over the seventeen-year life of the drug's patent. The chemical name of Viagra is 5-[2-ethoxy-5-(4-methyl piperazin-1-ylsulfonyl) phenyl]-1-methyl-3-pro-pyl-1,6-dihydro-7H-pyrazolo[4,3-]pyrimidin-7-one. That name is recognized only by well-educated chemists. Otherwise it has no wider use.

Obtaining the generic name is the first major objective of the company. Early in the process of drug development, the manufacturer submits up to three possible generic names to the US Adopted Names (USAN) Council, which selects one. In the case of Viagra, the approved generic name is sildenafil. In an attempt to organize generic drugs by category, the USAN Council has adopted approved stems that are attached to the end of the names of drugs that are therapeutically related. For instance, the stem in sildenafil is –fil, which is shared by its two competitors, Cialis (tadalafil) and Levitra (vardenafil).

Since the trade name is the name that will be widely used, it must pass a number of tests. The name cannot not indicate the product's intended use or imply its efficacy. The name must be simple, having no more than four syllables, so that people can readily pronounce it and easily remember it. The name must not conflict with any word that exists in English or in any foreign language where its meaning might be confusing. Furthermore, the name should not be one that, when used in ordinary patterns of speech, could easily alter the intended propose of the drug, such as to cause embarrassment or confusion.

There are still other requirements. No drug name destined for global use can begin with the letters H, J, K, or W, since in the languages of some 130 countries these letters do not exist, or they are pronounced differently than in English. The name cannot imply a cure and cannot apply to a specific part of the body, since many drugs first introduced in the market are later found to have other uses not originally anticipated. The last requirement is that the drug's name is unique and essentially meaningless. The names of these drugs are not supposed to directly reflect what they do, but the pharmaceutical companies like to come as close as they can in the hope that the name will be a big factor during the selection the consumer makes (consciously or unconsciously) when he purchases the drug.

After the drug company chooses a trade name, it submits the name to the FDA, which checks for potential look-alikes and/or sound-alike words. The FDA's Office of Postmarketing Drug Risk Assessment screens the name and has health-care professionals evaluate the results in simulating the ordering and writing of prescriptions for the drug. It also identifies other problems, such as variations in regional pronunciations of the name. The name Viagra passed all of the preceding tests and emerged as one of the most successful trade names ever developed.

There has been a lot of speculation about how the name Viagra was arrived at in the first place. The letter V is a favorite among experts in the field of naming drugs. V is one of the three letters most favored (the others being X and Z) because they are believed to have a dynamic and futuristic ring to them that makes them more marketable. Some believe that the name came about because of its relationship to the words *vigor* or *vigorous*, implying strength or vitality. Others saw an inference to Niagara Falls, epitomizing flow, power, and grandeur. Perhaps it was the combination of *Vi* from *vigor* and *agra* from *Niagra* (sic) to create one of the most successful global drug names ever conceived. It is a name that has become the hallmark for worldwide sales of billions of dollars throughout its life.

The names of the other two erectile dysfunction drugs are also of interest: Cialis and Levitra. Cialis is derived from the French word *ciel*, meaning "sky" or "heaven," implying that the drug takes the user to a heavenly place. It is interesting the Cialis is the market leader in France over its two other competitors. Levitra carries the strong implication of lever, leverage, or levitate. In passing, the *vitra* in Levitra contains the magic V, and some think it vaguely sounds a little like Viagra.

Erectile Dysfunction

Erectile dysfunction is the persistent inability to achieve or maintain an erection of the penis adequate enough for having satisfying sexual intercourse. This term replaces the older term of *impotence*, which was often regarded as pejorative, demeaning, or disgraceful. Erectile dysfunction, abbreviated ED, manifests itself in the following ways: (a) difficulty attaining or inability to achieve an erection, (b) erections that are too soft for penetration, (c) inability to maintain the erection after penetration, and/ or (d) an inability to maintain the erection long enough to achieve orgasm.

The introduction of the term *ED* owes much to the marketing campaigns for Viagra. If Viagra had been marketed as a drug for impotence, the drug would automatically carry the unpleasant inference of impotence. But marketing Viagra as a drug for ED gives the drug respect, particularly since the term *ED* entered the language as acceptable. One wonders how people looked at the word *dysfunction* (in erectile dysfunction) and either didn't understand its meaning or understood it to be a medical term and therefore acceptable.

Erectile dysfunction is very common. According to one study, as many as 30 percent of American men experience chronic ED. Erectile dysfunction affects the psyche as well as the body. The fear of no longer being able to have normal erections leads to a state where a man has doubts about his virility and his capacity as a lover, which only aggravates the problem. These factors have a serious impact on a couple's relationship. The man's partner may feel less attractive to him and even suspect him of being unfaithful. Many men avoid discussing the problem with their doctors or with their partners. The now widespread use of the term *ED* has made it easier to talk about it, since ED has not taken on the unmanly sense that was connected with the older term of *impotence.*

The Introduction of Viagra

In 1995, the diamond-shaped blue pill called Viagra as a treatment for ED was introduced. Viagra quickly gained universal name recognition and thus joined the ranks of such other widely recognized names as Coke or Nike. The initial release of Viagra catapulted sales to skyrocket proportions, which held steady for the next five years. Since the estimates of the incidence of ED were high, the news of its effectiveness spread widely and rapidly. Over the first five years of use, Viagra was the subject in 54,678 stories in print media. One of the popular hosts of a late-night talk show over the same period told 994 Viagra jokes. Two million hits were registered by one of the search engines on the Internet. Clever TV ads featuring ex-senator Bob Dole or the famous ex-soccer player Pelé captured the interest of the viewing public. These publically recognized figures talking openly and personally about ED convinced many ED sufferers that ED was not a taboo topic, but one about which they could talk to their doctors.

That the drug was very effective was more than hearsay when the results of multicenter, controlled trials in the US, UK, Europe, Japan, and Australia were published. From 1995 to 2005, twenty such trials consisting of 3,000 largely Caucasian men with ED were completed. Fourteen of the trials evaluated a single dose of sildenafil, while the others examined the effects of increasing the dose from low to moderate to high levels. More than 80 percent of the subjects completed the trials. The dropout rate due to adverse drug effects was only 5 percent.

In every trial of sildenafil, treatment caused improvements in erections when compared to the placebo. Two measures were used to assess improvement: penetration of the partner and maintenance of the erection. Overall, the sildenafil-treated men were four times more likely to experience improvement than those with the placebo. Furthermore, increasing the dose from 25 milligrams to 50 milligrams and to 100 milligrams resulted in an increasing number of positive responses. The drug did not produce desire for sexual activity but was effective once sexual activity was initiated. In other words,

Viagra is not an aphrodisiac. Viagra's most common side effects are head-ache, facial flushing, and upset stomach; less common (and brief) side effects are bluish vision, blurred vision, and sensitivity to light.

A New Use

Following the years of widespread use of sildenafil for ED, a new use for the drug was discovered for the treatment of pulmonary arterial hyperten-sion (PAH), i.e., high blood pressure in the arteries carrying blood to the lungs. Pulmonary hypertension (PHT) is a rare lung disorder in which the arteries that carry blood from the heart to the lungs become narrowed, requiring a higher pressure for blood to flow through the vessel, which results in pressures far above normal levels. The abnormally high pressure puts a strain on the right ventricle of the heart, causing it to expand in size. Over time, the overworked, enlarged right ventricle becomes weaker and diminishes its ability to pump enough blood to the lungs. When severe, this leads to failure of the right side of the heart. It is generally believed that the process starts with injury to the layer of cells that line the small blood vessels of the lungs. This injury, which occurs for unknown reasons, may cause changes in the way these cells interact with the smooth muscle cells in the vessel wall. As a result, the smooth muscle contracts more than normal and narrows the vessel.

Pulmonary hypertension occurs in individuals of all ages, races, and ethnic backgrounds, although it is much more common in young adults and is approximately twice as common in women as in men. Estimates are that only three hundred new cases are diagnosed each year. Symptoms of pulmonary hypertension do not usually appear until the condition has pro-gressed. The first symptom is shortness of breath with everyday activities, such as climbing stairs. Fatigue, dizziness, and fainting spells may occur. Swelling of the ankles, abdomen, or legs; bluish lips, skin, and/or nail beds; and chest pain may occur as strain on the right side of the heart increases. In more advanced stages of the disease, even minimal activity will pro-duce some of the symptoms. Other symptoms include irregular heartbeat, racing pulse, dizziness, progressive shortness of breath during exercise or activity, and difficulty breathing at rest. Eventually, it may become difficult to carry out any physical activities. Without treatment, the life expectancy of the victims is three years.

Since the blood vessels in the lungs contain PDE-5, sildenafil seemed a logical choice for treatment. Sildenafil ought to work through the same basic chemical mechanisms as in the case of erectile tissue of the penis, since it dilates the blood vessels of the penis to allow an increased blood flow to the penis. In the lung, the drug inhibits PDE-5 in these small blood vessels of the lung, resulting in prolonging the life of cGMP and a widening of these vessels. The overall result is an increasing ability of the heart to send blood to the lungs at a lower pressure. Lowering the blood pressure leads to improvement in the level of physical activity and of well-being.

In June 2005, the FDA approved the use of sildenafil for the treatment of this disorder. The approval was based on a well-designed study of 277 patients with PAH that demonstrated that doses of 20 milligrams, 60 milligrams, and 100 milligrams given three times a day reduced the level of blood pressure in the pulmonary arteries. Accordingly, the new name of Revatio was adopted and the drug adopted a new appearance of small round white pills, obviously very different from the blue diamond-shaped tablet of Viagra.

In May 2009, the FDA approved another new drug of the same family, called Adcirca, based on Cialis (tadalafil). It is an orange-colored, almond-shaped tablet, quite distinctive from Cialis's appearance. The recommended dosage schedule for Adcirca is once daily, while for Revatio, it is three times a day. The drug was well-tolerated with some side effects that were not thought to be serious. Since all three of the doses studied were equally effective, the final recommended dosage schedule was 20 milligrams three times each day. Of the 277 patients in the original studies, 259 patients continued to take the drug for a longer-term evaluation. At a one-year follow-up, 94 percent of this group were alive and had achieved further improvement in exercise tolerance. The most common side effects of Revatio were headache, indigestion, flushing, insomnia, and nosebleeds.

A Fascinating Discovery

First, a few words about the master clock of the brain, which is present in all mammals: The master clock is located in a highly specialized group of cells of the hypothalamus called the suprachiasmatic nuclei and known briefly as the SCN. The SCN receive information directly about illumination of the environment from specialized photosensitive cells of the retina, which set off a series of physiological events that generate wakefulness. In 2007, three Argentinian scientists, Drs. Patricia V. Agostino, Santiago A. Plano, and Diego A. Golombek, showed that the master clock that governs the circadian rhythms of the hamster contains cGMP as well as PDE—the critical elements underlying the action of Viagra on erectile tissue of the penis. (See the drawing presented in this chapter.) Using this information, they performed some fascinating studies on the light-dark cycles of the hamster.

They subjected hamsters to a daily routine of fourteen hours of light and ten hours of darkness. After a period of acclimatization, the researchers abruptly switched on the lights six hours early during the darkness period and continued the same light-to-dark routine from that point onward. Prior to changing the hamsters' routines, each animal received an injection of either sildenafil or saline solution. The scientists then measured the time it took the hamsters to restart their daily habit of running on a wheel. Hamsters receiving a large single dose of sildenafil resynchronized their body clocks and resumed their usual wheel routines within six days. Hamsters receiving a lower dose took eight days, whereas those receiving the inert injection took twelve days.

Many travelers find that the effects of jet lag are the strongest after an eastbound flight, which pushes the circadian clock forward. It is important to note sildenafil's circadian readjustments of the usual eastward orientation of the master clock. In other words, the drug simply advanced an innate rhythm. That suggests that the drug would work for eastbound travelers and airline personnel after long flights over a number of time zones.

This work was hailed by Improbable Research, a well-established group that collects and publishes the results of projects based on improbable research in many different fields. This group also awards annual so-called "igNobel Prizes" to honor achievements that first make people laugh, and then make them think. The prizes are intended to celebrate unusual research that honors the imagination and spurs people's interest in science, medicine, and technology. In 2007, it awarded its ig Nobel Prize in Aviation to the previously cited Argentinian scientists for their discovery that Viagra shortens jet lag recovery in hamsters. This work acquired great legitimacy by its publication in the prestigious journal *Proceedings of the National Academy of Sciences*.

REFERENCES

Agostino, P. V., S. A. Plano, and D. A. Golombek. "Sildenafil accelerates reentrainment of circadian rhythms after advancing light schedules." *Pros Nat'l Acad Sci.* 104 (2007), 934–939.

Burls, A., L. Gold, and C. Clark. "Systematic review of randomized controlled trials of sildenafil (Viagra®) in the treatment of male erectile dysfunction." *Br J Gen Pract* 51 (2001), 1004–1012.

Campbell, S. F. "Science, art and drug discovery: a personal perspective." *Clin Sci* 99 (2000), 255–260.

Osterloh, I. "How I Discovered Viagra." *Cosmos: Life Sciences.* July 18, 2007.

X-rays: Bertha's Hand

It's unearthly, it's truly mystical.
—Wilhelm Röentgen, 1895

Würtzburg, 1895

November the 8th of 1895 was a bleak, chilly day in the Bavarian town of Würzburg, home of the University of Würzburg, where Wilhelm Conrad Röentgen served as the rector and professor of physics. As rector, Röentgen was burdened with many academic and administrative affairs, but he still found time to give a few lectures and to conduct some research. Röentgen's typical day began with a group session with students in the morning, a customary, heavy noonday meal at home with Frau Röentgen and their niece, who lived with them, followed by a short nap and a return to his office to take care of university business, returning for supper.

As the sun was setting in the late afternoon, Röentgen hurried home to prepare for starting a new avenue of research. This fifty-year-old professor had done distinguished work on the physics of solids and liquids, but recently he had been intrigued by some of the experiments on the cathode ray tube that had been carried out by a colleague, Professor Phillip Lenard, at the University of Heidelberg. Lenard had been studying so-called cathode rays that were given off in a closed, nearly complete vacuum tube when negative charged electrical particles called electrons were discharged from the cathode that was mounted in the tube.

Under these conditions, the cathode rays in the tube glowed in vivid colors and patterns. Lenard had been able to create conditions in which the rays escaped an inch or so outside the glass tube and into room air. This fascinating finding piqued Röentgen's curiosity. He wanted to reproduce Lenard's results and to explore some of the details of the escaping rays by pursuing some original ideas that he had.

A Dramatic Experiment

For his experiments, Röentgen needed a completely darkened space, for which he chose the cellar of his well-kept house, where he had established a laboratory. The several small cellar windows had lightproof curtains, but to be absolutely sure of darkness, he decided that all of the experiments would have to take place after the sun had set, which at that time of the year was 4:30 p.m. He left instructions with the other members of the household that no one should open the cellar door, since any sudden light would destroy his dark-adapted vision.

Röentgen had obtained a vacuum tube of the type that Lenard had used. The air inside the tube had to be removed by the use of tedious, hand-operated device that Röentgen's laboratory assistant operated over the two-day period prior to any of the experiments. He also needed a source of alternating current, since the power to the house was direct current.

Before starting his experiments, Röentgen conceived of the idea of making a photosensitive plate using a piece of cardboard on which he had embedded fine phosphorescent crystals that would glow when the escaping rays hit it. He also thought to bring along two decks of playing cards to measure the penetration of the rays by seeing how many cards would be needed to block the rays between the source of the rays and the plate.

Before turning on the high voltage, Röentgen wrapped the entire outer surface of the tube with thin black paper so that the dim glow of the escaping rays would be more easily seen when the brighter light coming from the displays of light inside the tube were blocked out by the black wrapping.

He was now ready for the experiment. After the laboratory assistant, Röentgen extinguished all the lights, plunging the cellar into pitch-black darkness. It took a few minutes for his eyes to become dark-adapted. Even so, he had to search to locate the high-voltage switch. Increasing the voltage, he saw the cathode rays begin to appear outside the tube. He reached around for the phosphorescent plate that he had made, finally locating it at the other end of the bench where he had left it before turning out the lights. But when he picked up the plate, it was not dark, as he had expected, but rather it was showing a flickering faint green light.

The Mysterious Green Light

What had caused that? At first, he thought it might be coming from the electrical coil that was used to convert the current. But that wasn't it. Could it be a reflection in a mirror that was hanging close by? But there was no source for any reflection. It must be coming from the tube. To check that, he turned the high-voltage switch off and the green light disappeared. But turning the switch back on, it reappeared.

It certainly wasn't the escaping cathode rays, since they couldn't travel the distance of a meter between the tube and the plate at the end of the bench. Their range was only a few centimeters from the tube. Furthermore, the cathode rays were yellow, while the color of the light on the plate was definitely green.

He picked up the plate with its weak streak of green light and held it vertically. As he approached the tube, the light became a round green cloud, and as he moved even closer, it became a smaller but a much brighter spot. This mysterious ray was emerging through the glass wall of the tube, through the black paper cover, and through several meters in the room air. He stepped back and turned the plate around so that its cardboard backing now faced the tube. The light glowed as brightly as before. He realized that this had been a foolish idea in the first place, since if the rays could penetrate glass, they could certainly penetrate a thin piece of cardboard. But that conclusion prompted the question of what material could block the rays. He tried a few playing cards with no effect, then a whole deck and then two decks. The rays easily penetrated all of the cards. Groping in the dark, he located and tried a thick block of wood, but that didn't stop the rays. Neither did a thousand-page bound handbook that he pulled down from a bookshelf close by.

By this time, Röentgen realized that whatever he was seeing was something new. These were rays generated from the cathode tube, but they weren't cathode rays. They traveled in a straight line, and they could penetrate glass, two decks of cards, a thick piece of wood, and a thousand-page book.

A servant knocked on the cellar door to let him know that supper was ready. It seemed like a good time for a break to consider his next step. He said very little at the meal, eating hurriedly and abruptly excusing himself. He roamed the house, gathering pieces of several kinds of metals as well as bottles containing different liquids and various chemicals. Descending the staircase, he turned the device on and began the testing. Of all the objects he tried, only some of the metallic objects blocked the rays, and the best was a piece of lead pipe that was several millimeters thick.

Naming the New Rays

Turning on a light, Röentgen sat down with his notebook to record all that he had observed during the evening. These mysterious rays were not cathode rays nor were they ultraviolet rays, since it was well known that UV rays could be easily blocked with light materials. What should he name these "new rays"? Looking at the alphabet, the next letter following U and V, the letters for ultraviolet, was W, but he rejected it because its shape tended to resemble U and V. The next letter was X, and X fit the bill perfectly, since it was original and carried the connotation of an unknown.

Röentgen's next days were filled with many experiments, all aimed at defining more precisely the properties of the X-rays. He tested a whole host of materials to see which would block the rays completely, which only partially, and which not at all. Lightweight substances such as water, rubber, and aluminum sheets were no obstacles; heavier substances like concrete partially blocked the rays, but gold, silver, and lead stopped the X-rays completely.

Was his particular cathode ray tube unique in producing these X-rays? He tested several others and each produced X-rays. Could a magnet or a prism deflect the rays? Neither worked. But the magnet experiments proved that the rays were not charged electrically, unlike the cathode rays that bore negative charges. He became so immersed in his experiments that he began to skip supper, preferring to work uninterrupted well into the night.

A Great Breakthrough

By this time, Röentgen had learned a good deal about X-rays, but he needed a practical way to demonstrate them without having to use a phosphorescent screen in utter darkness. He toyed with the idea of taking a photograph of the plate, but almost immediately he realized that the photographic plate itself might be sensitive to the impact of the X-rays.

Simply turning the beam aimed at such a plate and developing it yielded a picture of the image of the X-ray. It was a great tool for all future work. Since the photographic plates were

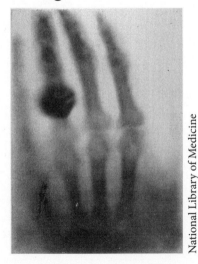

National Library of Medicine

covered with paper to block out the daylight, and since the X-rays easily penetrated the paper covering, the need to work in darkness was no longer necessary. He could precede working in broad daylight.

What would the X-ray pictures look like if the beam were focused on some ordinary object? For his first try, he used a weight used in a laboratory beam balance. The rays pictured the exact shape and size of the weight. Since Röentgen knew that the wooden box in which the weights were kept was nearly completely transparent to X-rays, he focused the beam on the weights inside the box. The result was a perfect picture of the weights themselves inside the slight shadow of the box. X-rays could enable one to photograph the contents of a closed container from the outside!

When he tested the piece of lead pipe, Röentgen had accidentally allowed his hand to drop in the path of the beam when he was positioning an object on the plate. When he developed the plate, he saw a shadow that had the exact shape of the pipe, but also the detail of the bone of two of his fingers. He was seeing his own skeleton!

Röentgen reasoned that an X-ray picture of a hand might be the most convincing way to illustrate his discovery to others. He couldn't use his own hand, because he couldn't pose his hand and operate the tube at the same time. Thus far, no one in or out of the household suspected that a great discovery had been made in the cellar. He decided to ask his wife, Bertha, to be the subject. He knew that she could keep his secret. Well aware of his intense activities in the cellar for many days, she agreed to cooperate. He led her down the cellar staircase, placed her hand on the photographic plate, and asked her to hold it still for six minutes. Then he asked her to wait while he developed the plate. It showed a beautifully detailed picture of each bone of Bertha's hand, including the shadow of the gold wedding ring on her third finger. When Bertha saw the plate, she was terrified. The idea of seeing her own skeleton provoked her crying out, "I have seen my death." She stormed up the steps, retreated to her bed, and covered her head with a blanket.

After working for six weeks, Röentgen realized that he must quickly publish his results in order to establish scientific priority for his discovery. He sent brief letter entitled "Ueber eine neue Art von Strahlen" (On a New Kind of Rays) to the Würzburg Physical-Medical Society with a plea that it be published as soon as possible in the society's proceedings. The editor agreed, and the published paper appeared on December 28, 1895.

Röentgen was aware that the society's published proceedings had a small audience and would not reach the much larger array of scientists who were working in the same or related fields of research. He therefore purchased a number of copies of his published paper and mailed them along with a letter to a large number of well-known physicists in Berlin,

Vienna, Paris, and London. Each letter also included X-ray photographs of the box of weights and of Bertha's hand.

The News Spreads

The idea that X-rays could see what had been hidden spread quickly and widely. Furthermore, anyone could make X-ray pictures because the apparatus was easy to operate, was affordable, and gave quick results. In those early days, X-ray pictures could be taken in "portrait" studios, in coin-operated machines, and at meetings of newly formed X-ray clubs. The following ditty became an instant hit:

> The Röentgen Rays,
> The Röentgen Rays
> What is this craze?
> The town's ablaze
> With the new phase
> Of X-ray's ways.
>
> I'm full of daze,
> Shock and amaze;
> For nowadays
> I hear they'll gaze
> Thro' cloak and gown
> —And even stays,
> Those naughty, naughty
> Röentgen Rays.

Röentgen suddenly became famous throughout the world. He refused to patent his discovery, believing that it should be used freely for the benefit of mankind. By 1901, Röentgen's discovery was universally lauded by the scientific world. He received many honors and awards as well as a long list of prizes, medals, honorary degrees, and memberships of learned societies. In several cities, streets were named after him. In spite of all this, Röentgen retained his modesty and reticence. In 1901, he was awarded the Nobel Prize in Physics "in recognition of the extraordinary services he has rendered by the discovery of the remarkable rays subsequently named after him." He accepted the award but gave no lecture, and he announced that the monetary award would be contributed to the University of Würzburg.

As the use of X-rays spread throughout the population, no attention was given to any danger. After all, the rays were invisible, and people who were being X-rayed suffered no immediate ill effects. No one conceived of the idea that X-rays would be harmful in large amounts over time. There were a few hints that too much X-ray could be harmful. For example, a man who had submitted to an X-ray of his head two weeks previously found that all of his hair had fallen out.

Criticisms and Attacks

Following his discovery of X-rays in 1895, Röentgen published three classic papers containing the details of his experiments. These were the last of his publications. Despite the widespread news of this discovery, Röentgen's personality did not permit him to celebrate his newly found fame. Indeed, he avoided any semblance of being a celebrated famous person. At the time, he was fifty years old and comfortably situated at the university. He had long been recognized as an eminent scholar of physics, and the year before, he was chosen as rector of the university, a great academic honor. His discovery of X-rays was widely hailed as "groundbreaking."

Yet voices of critics of his work were soon being heard. Otto Glasser, author of a magnificent biography of Röentgen, quotes a letter written by Röentgen to his friend Zander as follows: "My work has received recognition from many quarters. . . . This is worth a great deal to me, and I let those who are envious to chatter in peace; I am not concerned about that." But he *was* concerned, and every critical barb was a wound to his psyche.

Looking back, some critics questioned the originality and priority of Röentgen's discovery by citing earlier events that were probably due to X-rays, but were not further investigated. Sir William Crookes had observed that unopened wooden boxes of photographic plates that had been placed under his cathode ray tube showed shadows that were attributed to light leaks. Others had had similar experiences with X-ray plates, but had only concluded that it was advisable to store the plates some distance from the tubes.

Many early cathode-ray workers, such as Lenard, stated that they had observed a great number of new phenomena, but none were ever followed up. A Philadelphia physicist had accidentally made an X-ray picture over five years before Röentgen's discovery, but was unable to explain the phenomenon until Röentgen's observations were reported. Although most scientists gave Röentgen full credit and the honors that were his due, the rumor mill persisted with such items that the discovery was an accident or that the first crucial observation of the fluorescence of the screen was made by an assistant. Seasoned scientists of the day acknowledged the role of chance in Röentgen's work, but they also pointed out that only Röentgen had the gift of pursuing his discovery to unravel its basic cause.

Röentgen made no public responses to these charges, and over the period from 1895 to his death, he became an increasingly bitter man. One important effect of his bitterness was his refusal to publish anything further on X-rays after his three original communications. He also changed his will by stipulating that all correspondence, including some unopened papers about the discovery between 1895 and 1900, be burned after his death, a decision that regrettably was carried out.

Even though the news of discovery of X-rays was quickly widespread and often sensational, Röentgen assiduously avoided any self-publicity. He was very reticent by nature, and the many honors that he received were more of a burden than a pleasure. The University of Würzburg gave him the honorary degree of Doctor of Medicine, and he accepted honorary citizenship of his native town of Lennep. He declined all invitations to address scientific audiences on his discovery. Indeed, the only lecture he gave during his lifetime was the one he presented to the Würzburg Physical-Medical Society.

He declined the offer of nobility by the prince regent of Bavaria, who later bestowed the title "Excellency" on Röentgen. In 1901 he became the first Nobel laureate for physics and traveled to Stockholm to receive his award.

In 1900 Roentgen left Würzburg to take charge of the Physical Institute of the University of Munich, where he resumed his earlier work on the physical properties of crystals. After his retirement in 1920, he was given permission to use two rooms in the institute, where he continued to work until a few days before his death on February 10, 1923.

The Dangers of X-rays

An early alarm was sounded when Clarence Dally, Thomas Edison's chief laboratory assistant, suffered numerous radiations "burns" on his hands and face during his work on X-rays. A short rest away from the laboratory did not prevent recurrence of the burns, which developed into severe ulcerations and the development of a malignancy that required amputations of both of his arms, and that eventually led to his painful death. The experience was enough for Edison to abandon all X-ray projects and to never submit himself to X-rays for the rest of his long life.

Studies of the effects of X-rays on living cells in research laboratories were demonstrating the harmful effects on living cells and organisms. As early as 1929, Hermann Muller was awarded the Nobel Prize for his groundbreaking work that showed that X-rays produced mutations in the chromosomes of the fruit fly. These scientific advances plus the increasing cases of radiation sickness led to the development of safety regulations designed to protect both patients and personnel from harmful effects of X-rays. Even so, as late as the 1950s, many shoe stores still operated X-ray machines, which later were incriminated as causing testicular cancer in boys.

Comment

When Röentgen spotted that green light on the phosphorescent plate, he had the good sense to drop his original study of cathode rays and concentrated on unraveling the nature of these new rays. Röentgen's story is a wonderful example of an accidental discovery. In Röentgen's case, as in all other cases of chance discovery, the discoverer observes a completely unexpected event and stops whatever he was doing to pursue a new line of work aimed at understanding the significance of what he has just observed. Röentgen immediately grasped the significance of his discovery. He readily abandoned his study of cathode rays and launched a new investigation of the seemingly mysterious X-rays.

Röentgen's life is a sad tale of short-lived jubilation at his discovery, followed by a series of painful blows, however false, to his psyche, which resulted in a bitterness that persisted over the remaining twenty-eight years of his life. The world gave Röentgen its acclaim and acknowledged his discovery as one of the greatest in many decades, but in so doing, it made him a bitter man.

REFERENCES

Assmus, A. "Early History of X-Rays." 1995. http://www.slac.stanford.edu/pubs/beamline/25/2/25-2-assmus.pdf

Friedman, M., and G. W. Friedland. Chap. 6 in *Medicine's 10 Greatest Discoveries*. New Haven, CT: Yale University Press, 1998.

Glasser, O. "Strange Repercussions of Röentgen's Discovery of the X-rays." *Radiology* 45 (1945), 425–427.

Miller, A. "The History of the X-ray." Homepage.

Afterword: Acceptance of New Discoveries

Unthinking respect for authority is the greatest enemy of truth.
—Albert Einstein

Introduction

Looking back at the discoveries portrayed in this book, we may ask why it is that in some cases, acceptance by the general medical community was nearly immediate, while in others, it took so long for a discovery to gain wide acceptance. Discoveries in the first group include antibiotics, insulin, Viagra, Valium, ether, chloroform,[4] warfarin, and X-rays. In each of these instances, an evident need or the emergence of a new possible use, heretofore unrecognized, catalyzed the transfer of the discovery from the discoverer to the many thousands, even millions of people who benefited from it. By contrast, other discoveries, such as the artificial lens, digitalis, dental implants, the vitamins, chlorpromazine, lithium, and the antibiotic treatment of peptic ulcers generally took a much longer time before they were widely accepted.

---

4 Although chloroform is no longer used, the story of its discovery as an anesthetic is worth telling, particularly because it was the first successful attempt to ease the pains of childbirth.

Early Acceptance

With penicillin, it was the publicity in a London newspaper of the remarkable cure of a patient with bacterial meningitis that alerted doctors. Another, more extensive stimulus to the nearly immediate success of penicillin came from its treatment of soldiers' wounds during World War II that captured the attention of the world.

Selman Waksman, working with two Mayo Clinic doctors, Drs. Feldman and Hinshaw, launched the revolution of the first effective treatment of tuberculosis, and Jorgen Lehmann forged ahead with a companion anti-tuberculosis drug, para-amino-salicylate (PAS). The world had been waiting for a successful cure to tuberculosis, and streptomycin was it. With insulin, it was Banting and Best who dramatically pulled diabetic children and adults back from certain death. Their stories of survival led immediately to pleas from great numbers of diabetics for insulin. Karl Paul Link led the way for the acceptance of dicumarol and warfarin, especially following the great news event that President Dwight D. Eisenhower was one of the first patients to receive it.

With X-rays, there was Röentgen, whose discovery was widely aired in newspapers and in medical journals, who quickly showed the world the wonders of his discovery of the "new rays." Morton demonstrated that ether, and Simpson showed that chloroform, could induce sleep and introduced the world to pain-free surgery. Wells promoted nitrous oxide for painless dentistry.

By his discovery of cocaine anesthesia for the eye, Koller virtually immediately revolutionized eye surgery. The widespread immediate acceptance of Viagra and Valium occurred largely because of vigorous marketing campaigns by large drug companies, who saw and exploited the markets for these new drugs.

Thus the common denominator of the swift acceptance of all of the preceding discoveries occurred because each filled a previously large but unmet need. In many of the preceding instances, the drug became the first of a chemical family of modified drugs that followed. Newspapers and magazines added their voices by telling the stories of the medical miracles produced by the drugs. Lacking any significant opposition, each of these discoveries required no more than a year for complete acceptance by practicing doctors and the general public.

Delayed Acceptance

In contrast to the previous instances, a second group of equally important discoveries required a matter of several years before they were accepted into general medical practice. The fundamental problem of acceptance of this group is that it required a basic change in the paradigm of medical thought of the time. In other words, the prevalent medical thinking or mindset had to change 180 degrees before the discovery could be understood and accepted. Changing anyone's paradigm is difficult for most of us to do. Doctors may have an even greater problem, since their ingrained thinking was first embedded by professors in medical schools and amplified in their practices, which usually followed the leads of the most senior, opinion-making doctors who resided at the top of their various fields. These experts and their faithful apostles published papers, editorials, and opinion pieces in the leading medical journals that contained their persuasive opinions, which may have been deemed by many readers to be infallible.

A period of delayed acceptance followed the announcements of such new discoveries as lithium, the artificial lens, dental implants, chlorpromazine, vitamin B_1, vitamin D, digitalis, and the use of antibiotics for peptic ulcer. But in every case, an important advocate with convincing evidence was needed to present the value of the discovery, and a significant period of time was necessary to make a major change in the paradigm before the discovery received its proper widespread recognition and acceptance.

Many psychiatrists simply could not accept the idea that a simple chemical like lithium could radically change the behavior of manic patients. Since the field of molecular biology had not developed at the time, neither psychosis nor neurosis was thought to have any physical, much less chemical, basis. Lithium was condemned by Sir Aubrey Lewis and Michael Shepherd, two celebrated psychiatric authorities who wielded wide influence in the UK and in Europe in the post–World War II period. But their own opinions seemed so firmly set that neither of these experts had even tried lithium in their practices!

It took about twenty years for lithium treatment to become generally accepted. That goal was achieved because of the monumental and persistent efforts of the Danish psychiatrist Mogens Schou, who presented highly significant positive evidence derived from his well-designed clinical trials.

At mid-century, few if any British or American ophthalmologists could conceive of inserting an artificial lens into the eye of a patient. Their well-established paradigm was that foreign bodies in the eye should be removed, with the corollary that no foreign body should ever be inserted into the eye. Regardless of its merits, Harold Ridley's plastic lens was a foreign body, and all foreign bodies in the eye should be removed and never deliberately inserted. Furthermore, the great Sir Stewart Duke-Elder, the undisputed master of ophthalmology in the UK, rejected the idea of an artificial lens out of hand and simply refused to even examine the eyes of two of Ridley's patients, who had nearly perfect vision. Ridley did not give up. Over a long period, he persisted in implanting lenses with excellent results and continued to publish his findings, gaining an increasing number of less doctrinaire eye surgeons who had similar results.

With respect to dental implants, opposition arose early and was more or less sustained for a period as long as thirty years. The initial and most vocal was the Swedish dental profession, but by no means were they the only critics of Per-Ingvar Brånemark's work in patients. The crux of their opposition was the long-standing belief that any metal introduced in bone would end up as a bad result.

But a more fundamental problem of acceptance lay in the fact that Brånemark had discovered a unique biological phenomenon that he named osseointegration. Bone exposed to titanium results in a permanent fusion between the two surfaces in the absence of any intermediate tissue. For dental professionals, far removed from the basic sciences, this was difficult to accept, and there were some whispers that it was really a hoax.

Brånemark was steadfast as he continued to expand on, innovate, and publish useful applications of osseointegration. A major factor in widening the general acceptance of Brånemark's work was the positive results of a 1975 study by the National Board of Health and Welfare, the most important long-term consequence of which was widespread acceptance of implant procedures. A Canadian dentist, Dr. George Zarb, was instrumental in bringing Brånemark's dental innovations to North America.

Despite the excellent results of the use of chlorpromazine in trials of French institutionalized patients, American psychiatrists resisted its acceptance. Their traditional therapies were based on their particular paradigms that the treatment of psychosis could not possibly be achieved by the administration of such an innocent, basic chemical as chlorpromazine, in lieu of the then prevalent use of lobotomy and shock treatments, even though their results were often disappointing. They also objected to the trade name of Largactil, which led to a change of name to Thorazine. Dr. Heinz Lehmann, a leading Canadian psychiatrist, wrote a single published paper that led to the acceptance of chlorpromazine in North America.

Eijkman and Vorderman were met with an initial torrent of criticism over their rice experiments, mostly from the Dutch conservative medical establishment, whose paradigm was that beriberi was caused by an infection or a toxin and certainly not due to the lack of something in the diet. Fletcher's tightly controlled trial of human subjects deprived of the anti-neuritic principal (vitamin B_1), while ethically suspect, was successful in identifying the cause of beriberi. It took fifteen years from the time of Eijkman's first paper to the time of Fletcher's paper for a change in attitude about the essentiality of a dietary ingredient. Overwhelming evidence forced a very fundamental change in the paradigm about the cause of beriberi, an achievement that opened the door to understanding the vast field of dietary deficiency diseases.

Withering received heavy criticism of digitalis from John Coakley Lettsom, a London doctor with a willing audience for his views. To be sure, he had tried digitalis, but with a fatal result, probably because he selected the wrong patient and probably the wrong dose. For the next ten years, Withering used his passion for writing meticulous notes about each of his patients, which he summarized and published in a book that presented irrefutable results about the beneficial effects of foxglove.

Marshall and Warren's opponents were the vested interests of drug companies and the established practicing gastroenterologists, who followed the paradigm that ulcers were caused by a combination of stress, cigarette smoking, the extreme acidity of the stomach, and a certain reckless lifestyle. Furthermore, it was believed that no bacteria could survive in the acidity of the stomach. Marshall's risky self-experiment was the deciding factor in proving that *H. pylori* was the cause of peptic ulcer and gastritis, and that peptic ulcers could thus be treated by antibiotics. Over several decades of work, these researchers performed well-designed studies including a double-blind trial, the results of which showed that antibiotics cured peptic ulcers.

How were these problems of delayed acceptance solved? The simple answer is that when the results of definitive studies and observations were published, they were irrefutable. Changing an ingrained paradigm is certainly not easy, but it is made even more difficult by so-called experts touting their "beliefs" in the existing paradigm rather than championing a new one. Some of the examples cited earlier are exceptionally disingenuous when the "experts" failed to even test the proposed results for themselves. The situation is regrettable but made even worse when these experts at the height of their professions also play a dangerous role through their control of the leading medical journals' general medical practice. When the "experts" sense that they are being attacked by new facts, they behave as though the newcomers are dangerously undercutting their authority. Most of all they ought to know that sooner or later, they will fall from their self-appointed lofty pinnacles, as have many of their predecessors.

About the Author

Robert W. Winters is an experienced medical scientist, practitioner, and teacher. He is a graduate of Indiana University (AB summa cum laude) and of the Yale School of Medicine (MD cum laude). He did his postgraduate training in clinical medicine and in basic biochemistry at the Universities of California and Pennsylvania before accepting a Professorship of Pediatrics at Columbia University's College of Physicians and Surgeons in New York. The groundbreaking work that Dr. Winters and his research group accomplished in twenty years at Columbia is documented in 105 peer-reviewed papers and in four monographs. He received two national awards for excellence in clinical research and has been honored by his election to several of the preeminent American medical societies, as well as to the Royal Society of Medicine, London. During his tenure as professor, Dr. Winters was recognized as a physician with a critical eye and as a gifted teacher with a knack for explaining complex concepts in simple and understandable terms. His exceptional skills as a speaker have been recognized by medical students and house staff and by numerous guest lectureships, both in the United States and abroad. He now lives in Helsingor, Denmark, with his wife, and writes on medical topics of interest to the general reader.

Index